A Brave New
GLOBAL I
IN A CHANGIN

This new series of short, accessible
global issues of relevance to hum
enquiring reader and social activists in the North and
well as students, the books explain what is at stake and question
conventional ideas and policies. Drawn from many different parts
of the world, the series' authors pay particular attention to the
needs and interests of ordinary people, whether living in the rich
industrial or the developing countries. They all share a common
objective: to help stimulate new thinking and social action in the
opening years of the new century.

Global Issues in a Changing World is a joint initiative by Zed
Books in collaboration with a number of partner publishers and
non-governmental organizations around the world. By working
together, we intend to maximize the relevance and availability of
the books published in the series.

PARTICIPATING NGOS

Both ENDS, Amsterdam
Catholic Institute of International Relations, London
Cornerhouse, Sturminster Newton
Council on International and Public Affairs, New York
Dag Hammarskjöld Foundation, Uppsala
Development GAP, Washington DC
Focus on the Global South, Bangkok
Inter Pares, Ottawa
Third World Network, Penang
Third World Network–Africa, Accra
World Development Movement, London

About this Series

'Communities in the South are facing great difficulties in coping with global trends. I hope this brave new series will throw much-needed light on the issues ahead and help us choose the right options.'
Martin Khor, Director, Third World Network

'There is no more important campaign than our struggle to bring the global economy under democratic control. But the issues are fearsomely complex. This Global Issues Series is a valuable resource for the committed campaigner and the educated citizen.'
Barry Coates, Director, World Development Movement (WDM)

'Zed Books has long provided an inspiring list about the issues that touch and change people's lives. The *Global Issues for a New Century* is another dimension of Zed's fine record, allowing access to a range of subjects and authors that, to my knowledge, very few publishers have tried. I strongly recommend these new, powerful titles and this exciting series.'
John Pilger, author

'We are all part of a generation that actually has the means to eliminate extreme poverty world-wide. Our task is to harness the forces of globalisation for the benefit of working people, their families and their communities – that is our collective duty. The Global Issues series makes a powerful contribution to the global campaign for justice, sustainable and equitable development, and peaceful progress.'
Glenys Kinnock, MEP

About the Author

Riccardo Petrella is perhaps best known in the English-speaking world for his path-breaking book, *The Limits to Competition*. This book, widely seen as the 1990s counterpart to the Club of Rome's *Limits to Growth* report of the 1970s, was first published in 1995 while Riccardo Petrella was President of the Group of Lisbon. It soon became an intellectual bestseller and has been now translated and published in twelve languages.

Born in 1941, a social scientist by background, Riccardo Petrella became a leading member of Jacques Delors' team at the European Union as Head of Forecasting and Assessment in the field of Science and Technology (FAST) when it was trying to deepen the social content of the EU as a political project to which the ordinary person in the street could relate. He is committed to a vision in which the right to life for every human being rather than market optimization is the inspiration for a new global social contract which can provide renewed hope for the future.

Riccardo Petrella's new book is an important development of certain ideas first outlined in *The Limits to Competition*. He is currently teaching at the Catholic University of Louvain.

A GLOBAL ISSUES TITLE

THE
WATER
MANIFESTO

Arguments for a World Water Contract

Riccardo Petrella

Translated by Patrick Camiller

Foreword by Mario Soares

Zed Books
London and New York

University Press Ltd
Dhaka

White Lotus Co. Ltd
Bangkok

Fernwood Publishing Ltd
Halifax, Nova Scotia

David Philip
Cape Town

Books for Change
Bangalore

The Water Manifesto
was first published in 2001 by

In Bangladesh: The University Press Ltd, Red Crescent Building,
114 Motijheel C/A, PO Box 2611, Dhaka 1000

In Burma, Cambodia, Laos, Thailand and Vietnam:
White Lotus Co. Ltd, GPO Box 1141, Bangkok 10501, Thailand

In Canada: Fernwood Publishing Ltd, PO Box 9409, Station A,
Halifax, Nova Scotia, Canada B3K 5S3

In India: Books for Change, 28 Castle Street, Ashok Naggar,
Bangalore, 560025, India

In Southern Africa: David Philip Publishers (Pty Ltd),
208 Werdmuller Centre, Claremont 7735, South Africa

In the rest of the world:
Zed Books Ltd, 7 Cynthia Street, London N1 9JF, UK and
Room 400, 175 Fifth Avenue, New York, NY 10010, USA

Distributed in the USA exclusively by Palgrave, a division of
St Martin's Press, LLC, 175 Fifth Avenue, New York, NY 10010, USA

Copyright © Riccardo Petrella 2001
Cover design by Andrew Corbett
Set in 10/13 pt Monotype Bembo by Long House, Cumbria, UK
Printed and bound in the United Kingdom by Cox & Wyman, Reading

A catalogue record for this book is available from the British Library
US CIP data is available from the Library of Congress
Canadian CIP data is available from the National Library of Canada

ISBN 974 7534 739 Pb (South-East Asia)
ISBN 1 55266 055 9 Pb (Canada)
ISBN 0 86486 ?495 7 Pb (Southern Africa)
ISBN 1 85649 905 7 Hb (Zed Books)
ISBN 1 85649 906 5 Pb (Zed Books)

To the memory of my father,
a *maestro di banda*
in the days when municipal bands
still played an important social role
in the villages of southern Italy.

CONTENTS

List of Tables and Figures

ABBREVIATIONS

CONAIE	Ecuadorian Indigenous Nationalities Confederation
CSD	(United Nations) Commission on Sustainable Development
EAUDOC	Information and Documentation Service of the International Water Office
FAO	Food and Agriculture Organization
GATT	General Agreement on Tariffs and Trade
GWP	Global Water Partnership
HDI	Human Development Indicator
IBCSD	International Business Council for Sustainable Development
ICOLD	International Commission on Large Dams
ILO	International Labour Organization
IMF	International Monetary Fund
MAI	Multilateral Agreement on Investments
OECD	Organization for Economic Cooperation and Development
PI	Poverty Indicator
UNCED	United Nations Conference on the Environment and Development
UNDP	United Nations Development Programme
UNEP	United Nations Environment Programme
UNESCO	United Nations Educational, Social and Cultural Organization
UNICEF	United Nations Children's Fund
UNIDO	United Nations Industrial Development Organization
WHO	World Health Organization
WTO	World Trade Organization
WWC	World Water Council

ACKNOWLEDGEMENTS

I am first of all pleased to express my deep gratitude to Dr Mario Soares, who has believed unhesitatingly in the good cause of a World Water Contract ever since I first spoke with him of it in 1997. His personal involvement has been and still is a human and cultural asset of great importance for the success of the initiative.

Of the other public figures and friends who have believed in and made invaluable contributions to the drafting of *The Water Manifesto*, special mention should be made of: João Caraça (Lisbon); Vicente Perez Plaza (Valencia); Raúl Alfonsín (former president of the Republic of Argentina); Prince Laurent of Belgium (whose convictions and enthusiasm for the contract have been a source of tremendous encouragement); Larbi Bouguerra (Tunis); Susan George (Paris); Cándido Mendes (Brazilian senator); Rinaldo Bontempi (European deputy); Pierre-Marc Johnson (former prime minister of Quebec); Hasna Moudud (Dhaka); Abou Thiam (Dakar); Sunita Narain (Delhi); Pierre-Frédéric Tenière-Buchot (Paris); Driss Ben Sari (Rabat); Mario Albornoz (Buenos Aires); David Brubaker (Detroit) and José Antonio Pinto Monteiro (environment minister, Cape Verde); as well as Antonio Gonçalves Henriques (Instituto da Agua, Portugal); Mario Baptista Coelho (from the president's office, Republic of Portugal); Mario Lineo Correia (IPE–Agua de Portugal); Rafael Blanco Castany (member of the Valencian government); Madame Osita Bento Meneu and André Sobral Cordeiro (Mario Soares Foundation), remarkable linchpins of the two meetings in Lisbon of the committee for a World Water Contract.

FOREWORD

The Primacy of Politics

The role of politics

It is only in the last ten years or so that water has been one of the main issues on the political agenda, both nationally and internationally. Previously, except in the case of floods and other natural or industrial disasters (such as an accident resulting in contamination) or large-scale or prestigious symbolic events (such as the construction of a dam), water was usually regarded as a technical or economic issue. It was a field for chemists, hydrologists, legal experts, engineers, technical and administrative personnel in charge of systems for the pumping, collection, piping, distribution, purification and protection of water supplies.

A number of developments have changed this situation: the growing pollution of rivers, lakes and ground water, the rapid population increase in huge cities, soil erosion, desertification, conflict between farmers (irrigation accounts on average for 70 per cent of water use) and city dwellers (10 per cent), battles between regions of the same country with very different water needs and supply levels, as well as between various states and their neighbours (some 240 of the world's major basins are divided among two or more countries). We have seen heads of state (the French president, for example, in March 1998 in Paris) publicly display their concern and take the stage in support of a world water policy. We have stopped counting the number of inter-governmental conferences, ministerial meetings or declarations and conventions signed by the highest state representatives that have sought to develop a common vision and programmes for cooperation and coordination in the national, international and global management of water.

What can and should be done through politics?

Practically speaking, politics here refers to two closely linked but differing realities. On the one hand, there is the political actor, a naturally partisan individual, who is the bearer of an ideology and a value system, who upholds a certain conception of life, society and economy, and who seeks to gain acceptance for that conception by means of political mobilization and elections held under direct universal suffrage. He or she is thus a socialist, a liberal, a Christian democrat, a communist, and so on. On the other hand, there is the state, which constitutes the organized expression of politics defined and willed by the society of human beings. Although dependent upon the leanings and actions of politicians, state politics has the basic function of promoting and guaranteeing the general interest of a country's population as a whole, without discrimination on grounds of race, gender, religion, income, and so on. (Everyday reality demonstrates that this is not an easy task.)

Let me now answer the question contained in this section heading. As a politician, I am a socialist and I do my best to ensure that the socialist vision inspires water policy, both in my own country and at an international level. It is a vision based on the belief that water should be protected, developed, shared and utilized as a common good of humanity, and that priority should therefore be given to ensuring that everyone has access to it.

The state, for its part, should not only promote and ensure lasting and integrated management of the country's water resources (taking into account all the basic dimensions of a water policy). It should also, given water's special characteristic as a source of life and therefore a common good, adopt a global vision based on openness, solidarity and cooperation in its dealings with other countries, especially bordering countries that share the same sources of water.

A real state policy with regard to water will be capable of promoting sustainable development – in the sense intended by the 130 heads of state who signed Agenda 21 at Rio de Janeiro in June 1992 – so long as the policy is guided by principles of solidarity and mutual benefit.

This is why I am among the convinced sponsors of the World Water Contract, and why I am anxious that the foundation which bears my name should be closely involved in defining and implementing this contract.

I am especially glad to have the honour of presenting the *Water Manifesto* here in Lisbon, a city dear to my heart, on the splendid occasion of Expo '98.

The initial work that led to the idea of this manifesto was presented and discussed at Valencia, another great symbolic city with a rich and still living memory of a culture associated with water. For there in Valencia, the Water Tribunal – a popular, community-based organization – continues to exercise its judicial powers as it has done since 1492! This institution shows that the best solution to conflicts lies along the road of cooperation and respect for mutual interests.

Mario Soares

CHAPTER 1

ACHIEVING THE FIRST REVOLUTION OF THE TWENTY-FIRST CENTURY

The future for ourselves and our families – inhabitants of Brussels or Osaka as well as natives of Ecuador; Uzbeks, Tadziks and Turkmens of Central Asia as well as Americans of California – will depend less on technological and economic development than on the capacity of human societies to shape and administer certain rules, institutions and means of action that enable them to live together in an interdependent world with all its complexities and limitations, its diversity and its fragility.

West Europeans or North Americans hoped that they would not have had to pay for the effects of the South-East Asian financial and economic crisis, and that they would have escaped the fallout from the structural difficulties besetting Japan. This was a pure illusion. It is also an illusion to believe that the deforestation of sub-tropical countries, and the land degradation affecting Asia, Africa and Latin America, will continue to have only local consequences.

Furthermore, the heroes of today and tomorrow are not the most competitive, not those who succeed against the odds and conquer greater financial, commercial, technological and military power than others. The real heroes are those who advance the common good, who help assert the rights of all to life and citizenship.[1]

The great upheavals of the sixteenth, seventeenth and eighteenth centuries mainly revolved around systems of ownership, division and exploitation of the land, while the revolutions of the nineteenth and twentieth centuries were focused on the ownership, appropriation, distribution and exploitation of energy resources (coal, oil, electricity). Both gave rise to (or strengthened) systems of regulation that crystallized mainly around the emergence of new social classes and national states.[2]

In today's world, systems of regulation crystallize more and more either at the non-national level of multilateral global organizations that are destatized (the IMF, the World Bank, the World Trade Organization, etc.) or even privatized (the International Telecommunications Union and various other bodies responsible for norms and standards), or, in rarer cases, at the level of supranational state organizations such as the European Union.

This crystallization involves new social relations among social actors operating at an international and global level (most notably in finance, industry, agriculture and media), who are no longer identifiable with the social classes we have known. Rather, they reflect the interests, cultures and practices of new global classes currently in gestation, such as the 'class' of top managers and administrative personnel of industrial, financial and 'tertiary' multinationals.[3]

Today these new actors fight it out – or cooperate – mainly over the control of access to the basic resources which condition not only the lives of individuals but also the collective life of various regional, national and global communities. These basic resources are money, information and water.

The 'lords of the earth' are no longer industrial magnates like the old Rockefellers, Fords, Thyssens and Solvays, nor oil or railway 'barons'. They are, on the one hand, Bill Gates, Bertelsmann, Ted Turner, the Murdochs and Intel, and, on the other hand, financial corporations such as Morgan, Goldmann Sachs, Citibank, Fidelity and other investment trusts or insurance companies.

If present trends in relation to water continue for the next twenty to thirty years, the 'lords of the earth' threaten to become 'lords of water', the most likely and credible candidates being Suez–Lyonnaise des eaux, Vivendi (which includes the Compagnie générale des eaux), Saur–Bouygues, Nestlé, Bechtel, United Utilities, Danone, among others.[4]

More than a democratic reconquest
of information and financial disarmament

The upheavals presently under way and likely to follow in the early twenty-first century all have as their cause and substance the control of the three above-mentioned resources. The trends are clear enough in the case of money: growing financialization of the economy, primacy of finance over industry, liberalization of capital movements, floating exchange-rate systems, globalization of financial markets and their instability and volatility, abdication by politicians and political institutions, and delegation of monetary policy to the financial markets. Equally clear are the huge shifts that have affected and are currently reshaping the 'information society': industrial and financial concentration, liberalization and deregulation of the information markets, privatization of the information industry and services, shrinking of politics and the public space, primacy of technological innovation, tendency to standardization and homogenization of cultures, Americanization and supremacy of all things Anglo-American, explosion of the Internet and networking, virtualization of information, and so on.

Two major revolutions are needed over the next fifteen to twenty years to face up to these tendencies. The first would build on the huge potential for socio-economic change and innovation bound up with the new information and communication technologies, placing it at the service of a new drive to raise literacy levels and capacities for exchange and communication among the greatest possible number of people in each country. Instead of sharpening inequalities and the split between 'haves and knows' and 'have-nots and know-nots' – which is the case today and the most likely future trend[5] – such a revolution would lay the basis for the exercise of democratic citizenship by developing the opportunities for interactive freedom, plural autonomy and cooperation that the new remote information systems can theoretically offer.

The second would put an end to the unbridled financialization of the economy, to the supremacy of 'global financial markets' over national institutions of representative democracy and of politics in general. A revolution in this domain would rapidly shackle the power of finance over the next ten years, not out of hostility to finance as such but in order to restore its true role as an instrument for the promotion and development of the general welfare (rather than the opposite).[6]

The breakdown in October 1998 of negotiations toward a 'multilateral agreement on investments' (MAI – an attempt by the richest countries to give private capital complete global freedom of manoeuvre and decision making), as well as the fiasco of the WTO 'Millennium Round' talks in Seattle in December 1999 (designed to expand and strengthen mechanisms for the freeing of trade in goods, services and capital), demonstrate that a second revolution of this kind is not an impossibility. World civil society seems to have scored some important points on the way to a global future inspired by the principle of truly universal social welfare.

This said, it is becoming ever more evident that the first revolution of the twenty-first century, if there is to be one at all, will concern the rights of life and rights to life.

The basic meaning of the 'water revolution': a right to life for all

The purpose of change is to create the conditions for political action in our societies:

- to promote, consolidate and guarantee constant access to life for every human individual and community;

- to 'make society' together as a 'global human community', in the belief that guarantees of constant access to life are a joint responsibility of all human societies.

This revolution, then, involves a system of governance/regulation of the ownership, appropriation, distribution, management, protection, utilization and conservation of the principal source of life (together with air) for every living form in the earth's ecosystem: that is, water. As things stand, drinking water in particular is inaccessible to a large and growing number of people (more than 1.4 billion), and the ever greater pollution of surface and underground water, along with many other factors, does not encourage one to think that the future will be more favourable.

One can live without the Internet, without oil, even without an investment fund or a bank account. But – a banal point, though one regularly forgotten – it is not possible to live without water. This is why the water revolution is of such fundamental importance, more so than the revolutions necessary in information and finance.

The Water Manifesto is intended to demonstrate symbolically, politically and technically, on the broadest and most rigorous empirical basis, the urgent need for a 'water revolution'.

The first revolution of the twenty-first century involves the recognition:

- that, as a result of man's unrestrained and destabilizing intervention, we have reached a stage of life on earth when the future of various human communities and of the planetary ecosystem is part of a 'finite interdependent world';[7]

- that this future belongs to all of us, and will depend on our actions and interrelations;[8]

- that in this context we should guarantee access to life for every human being and living organism, by establishing at local and global level, and on a basis of solidarity, sustainable systems of ownership, distribution, management, use and conservation of the basic vital resources;

- that for this purpose it is necessary to begin at the beginning, by recognizing water as a common global heritage of humanity, as a source of life and a fundamental resource for sustainable development of the ecosystem Earth.

The following facts, among a host of other significant ones, well illustrate the urgent necessity of considering water as our common global heritage. In January 1998 the authorities of Papua New Guinea announced that more than a million people out of the country's population of 4.3 million were in a critical situation because of the drought-induced shortage of water and other necessities.[9] People still die for lack of water. According to the United Nations Development Programme, this is the fate of 15 million human beings every year.[10] Moreover, a few months later, in June 1998, Papua New Guinea was ravaged by floods that caused several thousand casualties – the same year in which 30 million people in Bangladesh underwent a huge crisis as two-thirds of the country lay under water for two months.

Also in June 1998, the international press reported violent protest demonstrations (burnt-out cars, looted shops, etc.) in the popular districts of Karachi, a city of more than 8 million inhabitants and Pakistan's chief port on the Indus Delta. The cause: failure to distribute water for two whole days – a common occurrence throughout the year.[11] It is normal and understandable that populations which have suffered permanent (or frequent) amputation of their right of access to a vital necessity should rebel in a forceful manner. As we said, there are more than 1.4 billion people who do not have access to drinking water – which equals 140 times the population of Belgium, or nearly five times the population of the United States. And more than another 2 billion have no system for domestic sanitation or the purification of waste water.[12] If nothing is done to reverse present trends, the number of people without access to drinking water in the year 2025 will rise above 4 billion, half the world's population.[13]

The new forms of exploitation and pollution of water supplies by industrial activities hitherto regarded as non-polluting (in this case the computer industry) are illustrated by a report in *Le Monde* that IBM pumps 2.7 million square metres of water per annum from the Neocomian strata beneath the French *département* of Essonne.[14] To produce its 64 megabyte microchips, the IBM factory needs very pure water such as one finds only in ancient reserves of this kind. But in this case they are protected by official provisions for water in the Seine-Normandy basin. So why did the authorities give the go-ahead for exploitation of this water table, rather than compel IBM to use (more expensive) surface water? According to environmental protection groups, the authorities are so desperate for opportunities of job maintenance or creation that they say and do very little against the powerful private multinationals. IBM is one of the largest employers in the computer industry in France, and its own position is quite simple: all its competitors tap the same ancient reserves of underground water, so it has to do the same to remain in the technological race.

This is an especially striking example of the great ambiguities and contradictions of our ways of dealing with the environment and sustainable development. Though recognized as crucially important, the requirements imposed by sustainable development often come into conflict with the priorities of corporate competitiveness. On the pretext of technological progress or with the lure of job creation (another key objective espoused by one and all), the imperative of competitiveness sweeps the board also in matters relating to water.

Another significant event, this time in 1997, was the transfer of responsibility for the Manila water supply from a public authority to two private-enterprise groups. Here in the capital of the Philippines, a city with a population of 10 million, the water situation is especially worrying: shortages affect more than 40 per cent of the population, more than 50 per cent of the water is lost through leaks, there is no purification system, and the available water is very often polluted. The first private group, which secured the rights for East Manila, comprises one Philippines company, the American Bechtel corporation and the British United Utilities; while the second group, which has been given West Manila, includes one Philippines company and, pre-eminently, the French Lyonnaise des eaux.

This transfer of responsibility was important for two reasons. First, it was the most important deal concerning water anywhere in the world, involving an investment of at least $7.5 billion, and it reinforced a growing trend for the ownership, appropriation, management and utilization of water in the major cities of under-developed countries to come under the control of private companies (other examples include Mexico City, Hanoi, Buenos Aires, Casablanca and Moscow). Second, Manila thus found applied to itself the principle according to which the poor pay for the water of the rich. For the first group proposed to charge Manila East (the richest part of the city where the business district is located) at a rate of 7 US cents per cubic metre, whereas the second group announced that it would sell water at 14 cents/m^3. Moreover, both rates were well below the price formerly paid by wealthy residents of Manila (approximately 28 cents/m^3).

Until recently (even in the United States), water remained one of the last goods and services to be touched by the wave of privatizations that has swept away nearly every public sector over the past twenty years: from public banks and insurance companies, through gas, electricity and railways, to postal services, telecommunications and hospitals.

Privatization is now changing the general shape of water services and no longer seems to elicit clear opposition among political leaders throughout the world. Doubtless this is the fruit of a decade of strong pressure exerted on these leaders and on public opinion to consider water mainly as an economic asset whose value, ownership and use cannot escape the laws of the market. We shall return to these questions in greater detail below.

A final event should be mentioned as especially significant for the coming years: namely, the agreement that put an end to the water war in the western United States, which had been raging for decades, between California and the neighbouring areas of lower Colorado, Arizona and Nevada. Under political and financial pressure from Washington, the local authorities in Los Angeles have now accepted (after forty years of refusing to do so) that they are responsible for excessive depletion of water from Lake Owen and for the devastating effects which this has produced on the lake and the surrounding region. They have also recognized – together with big farmers – their responsibility in relation to the Colorado river (one of the few in the world that no longer carries water to the sea). The agreement among all the concerned parties (inhabitants of the Lake Owen area, Los Angeles and San Francisco, big farmers, residents of San Diego, owners of the 560,000 Californian swimming pools that have drained fresh water from the mountains of Nevada, the populations of Arizona and Nevada) should finally permit integrated long-term management of the region's water supplies.[15] What this shows is, first, that economic growth and the satisfaction of corporatist interests (in terms of profit, lifestyle and consumption)[16] inevitably lead to conflict; but also, second, that conflicts relating to water can be resolved through an approach based upon common rules of distribution and use instead of trials of strength resulting in victory for some over others. This should be a source of inspiration for the authorities of all the many countries around the world that are engaged in conflicts over water (to name but a few: Turkey and the other countries through which the Tigris and Euphrates pass; Israel and its immediate Arab neighbours; the eight countries affected by the Nile basin; and the Mekong countries of Thailand, Vietnam, Cambodia and Laos).

All the lessons and prescriptions resulting from the five cases we have just considered are highly instructive. There is no valid technological, financial, economic, cultural, political or religious reason why human societies should allow water to become a rising source of conflict, disease, death, environmental destruction, urban degradation and social friction, when the facts themselves show that water can transform the quest for economic welfare and interest satisfaction into an opportunity for cooperation and joint development within a system of regulation that treats water as a common asset.

For this to happen, many rules and situations will have to change. This is the sense in which we may speak of 'the first significant revolution of the twenty-first century'.

We don't have time to wait until the year 2030

In order to escape from present trends, at least three changes will have to take place over the next ten to fifteen years.

The first change concerns the way in which we conceive of water, especially of relations between human beings and water within the framework of the relations between human beings and the ecosystem Earth.

In the last twenty years, the various visions of our societies (with regard to technology, sport, art or whatever) have become increasingly imbued with a techno-economistic culture. Most living beings and most of the earth's resources, even including the basic constituents of the human body, tend to be regarded as thing-like commodities fully translatable into economic values (costs, prices, profitability, productivity). Thus, especially since the mid-1980s, multilateral international bodies such as the OECD, the World Bank or the IMF – as well as associated agencies such as the Asian Development Bank or the Latin American Development Bank – have begun to speak more and more often and more and more clearly of water as an economic asset. Dozens of reports and conferences, and official declarations distributed all over the world, have endowed this 'vision' with a high degree of political legitimacy and scientific and economic credibility.[17]

The great majority of people from the world of economics (industrialists, financiers, insurance people and so on) have encouraged and supported this approach. It has not been a deliberate conspiracy. One is forced to conclude, however, that over the past decade or so there has been a great convergence or affinity between the visions and prescriptions that the major multilateral agencies and the business world have had in relation to water. This is the case, to take but one example among so many, with the positions advocated by the International Business Council for Sustainable Development (IBCSD) and by the World Bank.[18]

The currently fashionable notion of public/private partnership is certainly one to be fostered, resulting as it does from a justified desire to promote collaboration among all relevant actors by overcoming cut-and-dried positions (public on one side, private on the other). It must be said, however, that at present such partnership tends to be expressed in an unequal relationship whereby the role of the public is devalued (strategically, financially and operationally) and that of the private glorified. Analyses and proposals referring to the public domain all focus on its limitations and on the urgent need for deep reforms so that it can become an effective and credible partner. But when it comes to the private sector, the talk is mainly of the benefits to be gained from its involvement in hitherto public areas, and of the most effective ways of encouraging and facilitating its partnership with the public sector. Private is taken to be synonymous with efficient, profitable, transparent, flexible, adaptable and innovative.[19]

Public/private partnership in relation to water tends to cultivate and implement the visions and approaches of the private sector, so that water (the source of life) is in danger of gradually becoming one of the principal sources of profit, one of the last areas to be conquered for the private accumulation of capital.

Not every society has yet reached that point; some are even backtracking – witness the decision of the Netherlands government in late September 1999 to inform the Senate that water services will remain in public hands. A significant number of industrialists and economic associations are rightly aware of the importance of a genuine, balanced partnership, in which each side plays the role appropriate to it.

Water must be prevented from going the way of oil. To free the perception of water from the grip of techno-economistic concepts, to assert a vision of it not as *res nullius* (nobody's thing) but as *res publica* (a public good), as the first global *res publica* of societies calling themselves technologically and economically global: this is the first necessary change, and one that will certainly be difficult to achieve! Indispensable to it will be the mobilization of NGOs and trade unions, and scientific-political commitment on the part of both local and global intellectuals (especially in the media and the creative professions).

The second change concerns the state's appropriation of sovereignty and ownership rights over water. The statization of water, which accompanied and followed the periods of formation and consolidation of national states from the sixteenth century onward, played a historical role of the greatest importance. It made it possible to promote and safeguard a modicum of decent living conditions, and to make various attempts (rather weak until the end of the nineteenth century) to reduce inequalities between rich and poor in access to basic goods.[20]

The other side of the coin was also important. One thinks, in particular, of:

- the bureaucratic-statist centralization of decision-making powers with regard to the exploitation, utilization and management of the country's water resources;[21]

- the emergence of policies for the development of water in the service of nationalist, or even expansionist, geopolitical and military strategies: it is well known that major works for the harnessing of rivers or the construction of dams have been and still are dictated by considerations of regional power and positioning – a 'state logic' at the origin of numerous conflicts over water;[22]

- corrupt practices between the central/local state, political organizations and the business world in the awarding of concessions and indirect or joint management contracts for various water services.

It is important to destatize water: that is, to free it from the bureaucratic-centralist logic of state power by affirming the value of state citizenship. Destatization of water does not, however, mean privatization in the form of a transfer of ownership and control to private corporations. A cooperative type of enterprise delegated to run a public service (one that really does operate on the basis of cooperative principles) is neither the state nor a private capitalist company; the same is true of certain social services provided by private organizations of so-called civil society. The destatization of water may therefore take forms that have very little in common with privatization in the sense of the term that currently has such widespread currency. In India, for example (though the same holds in many African and Latin American countries), the destatization of water and other public goods has a mainly positive and progressive connotation, for it would involve the transfer of ownership rights and managerial powers to local communities (especially villages), in accordance not only with an age-old precolonial tradition, but also with the real necessities of Indian society today.[23]

We should not, of course, idealize village communities or grassroots urban communities. In some countries, community or municipal ownership of water has actually weakened the capacities of public authorities, as it is not easy for local collectives (even in big cities) to stand up to powerful multinational corporations. Besides, the history of Europe includes a period that was marked by constant deadly conflicts over territory and power among a multitude of free and sovereign communes and municipalities. The small (or the ancient) is not necessarily more peaceful, solidaristic or egalitarian than the large (or the modern). The sterile battles for funds that oppose small NGOs to one another are a case in point. And in the last few years we have seen conflicts over water frequently erupt among small indigenous communities in Ecuador. Nevertheless, history also shows that a solid and lasting system of regulation is more likely to be constructed when the ones mainly responsible for water are the grassroots human communities. For they, more than large political entities such as nation states, tend to treat water as a common good.

In the present day, the destatization of water as defined above involves a new system of regulation and control which, in keeping with the principles, rules and framework of the World Water Contract (of which more below), entrusts the integrated management of water to public bodies such as local communities, citizens' groups, village or town networks, and cooperative societies.

The third change concerns the logic – inimical to solidarity and sustainability alike – that currently prevails in the global organization of agriculture and the growth of cities in Africa, Asia and Latin America.

Until recent times the crystallization of regulatory systems around the nation, and around provisional compromises based on the relationship of forces between the social classes, made it possible to affirm as a positive trend the transformation of peasant agriculture into intensive industrial agriculture. The aim was to ensure the country's food independence, while raising the farmers' living standards through general modernization of the national system of production, distribution and marketing of the food and non-food produce of agriculture. Similarly, the growth of cities in the West was regarded as a sign of a country's progress and power. Not long ago, people were proud to learn that London or New York was the most populous or most extensive city in the world. The planners of Paris 2000, working in the early 1980s, set themselves the goal of making it Europe's largest urban region, with a population of 12 million.

Today, world agricultural overproduction (the 'agriculture of surpluses'), together with commercial and industrial battles among the big American, European, Canadian and Australian producers, agribusiness and the major continental and global distribution networks, have made it abundantly clear that national solidarity (or European solidarity in the case of the Old World) is no longer more than a secondary factor. Indeed it is often a mere cover for the assertion of the real financial, industrial and commercial interests, and for increasingly global power strategies within a capitalist market economy based on intensive industrial methods.[24] We are familiar with the ecological disasters caused by such agriculture: the contamination and degradation of the land, and the pollution of surface and underground water by nitrates, nitrogen compounds, heavy metals and other toxic substances;[25] and its extension to poorer countries elsewhere in the world has led to the destruction of local agriculture and life systems. So-called 'developed' agriculture has long been in the dock almost everywhere, held responsible for allowing factors contrary to solidarity to become the driving force behind it.

Here are a few examples:

- While food production has never stopped growing, world hunger has not declined but actually advanced in recent years in some parts of Asia and, above all, Africa.[26] Moreover, a crisis is beginning to appear in the production of staples such as wheat and rice, as their rate of growth tends to decline to the advantage of foodstuffs with higher value-added (pet food, fast food, luxury items, products geared to consumers with high purchasing power).

• Land degradation is one of the most striking expressions of a logic different from and contrary to principles of solidarity. If land degradation continues, it is because those involved want it to (with a mass of pretexts such as: 'We can't change anything by ourselves', or 'If others don't change too, we'll be eliminated from the market', or 'Solutions will be found sooner or later – you have to trust in technology and innovation').[27] But the solutions are well known, in fact, as are the ways of reinventing an efficient and sustainable peasant agriculture, both locally and at the level of the world's major regions.[28]

The same considerations apply to the now destructive growth of big cities in the poorer countries of the world (the 'megapolises of megapoverty'). Between 1950 and 1990, the number of cities with more than a million inhabitants rose from 78 to 290; it will reach 650 by the year 2025. The great majority of these cities (250 or thereabouts) are in Asia, Latin America and Africa. And of the 21 cities that will very soon have a population of more than 10 million (15 of them located by the sea), 17 will be in the Third World.[29]

The social, economic and environmental sustainability of these cities is close to nil. One can certainly know happiness in Calcutta, but Calcutta remains one of the megapolises of megapoverty. Few are the people outside its natural catchment area who would choose to go and live there – apart perhaps from a tiny well-to-do minority for whom it might be a 'fascinating experience'.

Water made it possible to build cities; its shortage and misuse are robbing them of their future.

Not many people in the countries of the North are aware that the quality of water in our cities is declining as a result of the pollution and contamination of reserves. If this is not evident, it is because the ever-rising expenditure needed to clear used water and to maintain a pure supply is passed on to relatively affluent urban populations. Even an average increase of 50 per cent on an initial price of US$1.50/m^3 represents a relatively minor cost, especially when consumers are used to buying bottled mineral water at a price 500 or 1000 times higher than that of tap water.

City dwellers in the South, on the other hand, are all too well aware of the deplorable state of things, but for various reasons of a political, social and cultural nature the local ruling classes prove incapable of breaking the vicious circle of poverty and the local populations themselves seem resigned to their fate. Nevertheless, the conditions of hygiene and sanitation in their cities are catastrophic. In India, for example, 70 per cent of the population has no proper drainage system. Thirty to forty per cent of the population of Mexico City, Karachi, Manila, Jakarta, Rio de Janeiro, Buenos Aires, Casablanca, Delhi, Hanoi, Cairo, Shanghai and Seoul do not have access to drinking water. Those who can afford it drink bottled water. The massive and extremely rapid increase in the size of the population has a lot to do with this. For example, Mexico City went from 1.5 million inhabitants in 1940 to 15 million in 1990, Shanghai from 5 million in 1950 to 14 million in 1990, and Jakarta from 1 million in 1930 to a current total of 20 million for Greater Jakarta. By now Mexico City is also thought to be close to 20 million; Calcutta is already there, soon to be joined by Bombay.

Health is thus the number one problem for these cities.[30] Even when people have access to water, its quality is well below World Health Organization standards for drinking water. The case of arsenic contamination in the cities of Bangladesh and West Bengal is here particularly instructive – as well as deeply disturbing.[31] Nor is this all. The more that urban needs rise in Africa, Latin America and Asia for quantity and quality in the water supply, the less investment will be available for agriculture (especially for conversion to different crops). This will force city dwellers to spend more on importing food produce, which will in turn make them poorer and make it more difficult to find the sums necessary for infrastructural investment in water services, agriculture, health and education. The vicious circle of poverty will once more be closed.

Given the demographic trends and the pressures from an external socio-economic environment dominated by the game rules and value systems of the richer countries, the water needs of the major countries and cities of megapoverty will increase exponentially to impossible levels. The World Bank estimates that, if regions with a high rate of urbanization are even to maintain their present conditions of water supply and drainage (which are usually deplorable enough), they will have to invest one per cent of their GDP between now and the year 2025. For the so-called developing countries as a whole, the infrastructural investment considered indispensable for water projects over the next ten years will total 600 billion dollars – only 60 billion of which could, under present policy options, come from international sources. The conclusions are not hard to draw.

The changes analysed above will not be easy to achieve. The scale of the problems and the means needed to solve them (above all, the nature of the required collective choices and behaviour) might suggest that the 'battle for water'[32] stands virtually no chance of success.[33] Many consider that it will not be possible by 2020 or 2025 to halt the degradation of living conditions in the large cities of Asia, Africa and Latin America, where the population has neither the financial means nor the technology nor the skills which leaders and experts from the developed countries deem necessary to check present trends and to trigger dynamics of sustainability and solidarity. The same leaders and experts have given up the idea of assistance in their aid and cooperation policy (which in a way is not such a bad thing, given the strings that have previously been attached to it). The slogan of the day is now: 'Forget aid, think business!'

In this context, it is expected that the situation will worsen but that at least a small number of social groups, city areas and regions will manage to pull through the next ten to fifteen years. The hope is that these will be capable of giving rise to a new cycle of development, as sustainable as possible and affecting the greatest possible number of people.

The fundamental questions

The three fundamental questions are therefore the following:

- Why haven't we been able to lessen the scale of the water crisis in the world, despite the number of major national and international initiatives taken over the past twenty years, with considerable investment and the involvement of thousands of NGOs?

- Is there any hope that, if the control of water is left to depend on the political orientations current in the richer countries and on the funds available to the poorer ones, a way will be found to prevent more than 3 billion human beings (out of a total of 8 billion) from being without access to the source of life in 2020 or 2025, as has been predicted?

- Why should a pensioner in Quebec, a Volvo worker in Stockholm or a taxi driver in Neuchâtel not only feel concerned about the water problems of a poor peasant in Senegal, a resident of Calcutta or an unemployed person in Mexico City, but also be ready and willing to do something to ensure that every human being has access to safe drinking water and that every human community can use water for agricultural, industrial or other purposes?

The second and third of this set will be considered later. Here we shall analyse only the first question, a perfectly justified one to ask.

In fact, at least since the second half of the 1970s, and especially since the first major world conference on water (organized in 1977 by the United Nations at Mar del Plata, Argentina), world leaders have been aware of the scale of the problems concerning access to water of sufficient quantity and quality, and of the risks associated with growing shortages and degradation of the supply. The Mar del Plata conference set out the basic facts and made water one of the top issues on the international political agenda. And yet the 'water crisis' has continued to worsen, so that twenty years later another UN agency, UNESCO, could organize the umpteenth world conference under the title: 'Water: a Looming Crisis?'

Between the dawning awareness of a crisis in 1977 and the talk of a looming crisis in 1998 (the question mark is purely rhetorical) there was no end of conferences, congresses and forums – international, continental and global; sponsored by public authorities and specialist organizations in the field (see the list on pp. 24–5). Most of these events, which had an importance it would be superficial to minimize, gave rise to action programmes, projects, resolutions and declarations, some of them important moments not just in terms of awareness but also for the definition of new concepts (such as the right to water) and new solutions.

Appendix 1 and Appendix 2 list respectively the international or world professional organizations operating in the field of water, and the specialist agencies of the United Nations directly or indirectly involved in such issues.

In one way or another, all this high society actively participated in the International Drinking Water and Sanitation Decade, launched by the United Nations in the 1980s following the Mar del Plata conference to enable all men and women to have access to safe drinking water by the year 2000. But the results fell far short of this target. According to United Nations officials, the 'water decade' opened up access to water for an extra 600 to 800 million people. The key question, then, is why so little progress was made towards a solution of the problems.

The water decade was followed, after all, by an intensification of meetings, conferences and forums. Their frequency and importance were linked to the UN Conference on the Environment and Development (UNCED) held in June 1992 in Rio de Janeiro (which enshrined the concept of sustainable development and laid the basis for a world environmental policy) and to the resulting institution of a World Water Day on the 22nd of March each year.

The Rio conference did indeed help to reaffirm, within the framework of Agenda 21, the urgent need for a world water policy. A few years earlier, the World Bank had for its part proposed and sought to implement a concept of 'sustainable integrated management of water resources', and it is still continuing to do this.[34]

The United Nations and World Bank machines do not stop there. The UN Commission on Sustainable Development (CSD), created to follow through on the decisions, resolutions and agreements approved at the Rio conference, has become a locus of discussions and meetings on the water-related issues that are increasingly one of its chief concerns. It was in the run-up to the sixth session of the CSD, in late April and early May 1998, that two important world conferences took place on water, one organized by the German government, the other by the French.

In 1996 the World Bank, for its part, teamed up with various UN agencies, states (Sweden and the Netherlands, for example) and private corporations (Suez–Lyonnaise des eaux) to organize two major initiatives in this field: the foundation of the World Water Council (WWC), and the launching of Global Water Partnership (GWP) in Stockholm. The task of the WWC is to develop, propose and promote a common world vision on water-related issues.[35] As for the GWP, its aim is to get public bodies and private companies to work together on a water-saving resource policy

Main world conferences on water in 1997–2000

1997	Location	Event
11 March	Marrakesh	First World Water Forum
1–5 September	Montreal	Ninth World Congress of International Water Resources Association on 'Prospects for Water Resources in the Twenty-first Century: Conflicts and Opportunity'
3–7 November	Manila	Fourth Global Forum of the Water Supply and Sanitation Collaborative Council
November	Yokohama	Meeting of Public Service International, which approved a *Water Services Code*
18–20 December	Valencia	UNESCO-supported world conference on 'Water Management in the Twenty-first Century: Towards an International Tribunal'

1998		
27–30 January	Harare	Meeting of experts on 'Strategic Approaches to Freshwater Management', in preparation for Sixth Session of UN Commission on Sustainable Development
March	Bonn	International Conference on International River Basin Management, held on the initiative of the German government
19–21 March	Paris	International Conference on Sustainable Development and Water Resources, held on the initiative of the French government
15 April–1 May	New York	Sixth Session of the Commission on Sustainable Development (CSD), to apply Agenda 21 in the protection of water resources (Action Water 21)
3 June	Paris	International UNESCO Conference on world water resources: 'Water: A Looming Crisis?'
18–20 June	Lebanon	International Conference on 'International Law and Comparative Law Relating to International Watercourses: Education in a Culture of Shared and Protected Water', sponsored by the International University

1999

| August | Stockholm | Ninth Stockholm Water Symposium organized by the Stockholm International Water Institute, in particular the seminar on 'Hydrosolidarity' |
| 18–24 September | Buenos Aires | World Water Congress, organized by International Water Resources Association (IWRA) |

2000

| 13–15 March | Melbourne | Tenth World Water Congress, IWRA, on 'Water Management' |
| 17–22 March | The Hague | Second World Water Forum |

Source: Group of Lisbon

Main Declarations Concerning Water (1990s)

Nature of declaration	Context
'Montreal Charter' on *Water and Sanitation*	International NGO forum in Montreal, 18–20 June 1990, organized by Oxfam and others before the official closing of the International Drinking Water and Sanitation Decade
'Dublin Declaration' on *Water in a Perspective of Sustainable Development*	International Conference on Water and Environment, 26–31 January 1992, organized by United Nations in preparation for the UNCED conference in Rio de Janeiro in June of the same year
'Strasbourg Declaration' on *Water as Source of Citizenship, Peace and Regional Development*	European Forum, 12–14 February, organized by the International Water Secretariat, the Parliamentary Assembly of the Council of Europe, and Europe Water Solidarity
'Paris Declaration' on *Water and Sustainable Development*	International Conference on Water and Sustainable Development, 19–21 March 1998, organized by the French government in preparation for the Sixth Session of the United Nations CSD
The Hague Declaration on Water Security	Second World Water Forum, The Hague, March 2000

that will bring (soluble) demand back in line with actual supply.[36]

In January 1999 the WWC, in cooperation with most UN agencies and with the aid of a number of governments (especially the Netherlands) and the World Bank, created the World Commission on Water in the Twenty-first Century. The Commission was to develop and implement a Long-term Vision for World Water, Life and Environment in the Twenty-first Century on the basis of work carried out by the WWC. The Long-term Vision was presented and discussed at the Second World Water Forum Conference between 17 and 22 March in The Hague.

According to this basic document, the broad emphases of a world water policy to ensure that everyone has access to safe water should be as follows:

- Water is a scarce resource, a vital economic and social asset. Like oil or any other natural resource, it must be brought under market laws and opened up to free competition.

- Rational and efficient management of water resources requires a rigorous economic culture and practice. Water service providers, whether public or private, should set themselves performance targets to be measured by the criterion of consumer satisfaction.

- Water is the primary factor in health. A rational and efficient water policy should aim to achieve and maintain the best possible quality, and to this end more and more infrastructural and maintenance investment will be necessary around the world. Such huge sums can be guaranteed only by the capital market in accordance with the aim of profitability. Ultimately, therefore, water policy is a question of finance (access to investment, able to make a profit).

The institution of World Water Day has helped to launch or strengthen a range of activities (information, training, campaigning, protests, proposals) on the part of thousands of national and international voluntary NGOs which, even before the Mar del Plata conference, were actively engaged on water issues. Some of these organizations have an international reputation and are well covered by the media: for example, Greenpeace, Worldwide Fund for Nature, *The Ecologist*, Friends of the Earth, Earth, Enda, Green, Solidarité/Eau, Swissaid, Caritas and Oxfam.

For at least twenty years, then, millions of people have been working on and for water. Tens of thousands of leading politicians, scientists, economists and representatives of civil society have been fighting for the efficient management of water resources and to ensure that people have the 'right' access to it. Hundreds of programmes, projects and declarations have been approved, applied and implemented. Tens of billions of dollars, in addition to local expenditure, have been allocated and invested.

Yet despite all this, 'a crisis is looming' and there is still talk of a 'water bomb' about to explode.[37]

The incomplete explanation of the 'water crisis' (1.4 billion people having no access to drinkable water)

Many different reasons have been given to explain (though hardly to justify) this situation. They may roughly be divided into four groups according to the particular aspect they highlight:

- Unequal distribution of water resources
- All factors relating to the waste and mismanagement of available resources
- A worsening context of pollution and contamination
- Population growth, especially in Third World countries

The first category – the great inequality in the distribution of water resources – involves grave *local* scarcity in certain countries and regions. The threshold at which drinking water scarcity

begins has been set at 1,000 cubic metres per person per year, or
an average of 2,740 litres a day. Below 500 cubic metres the
situation becomes critical, while between 1,000 and 2,000 cubic
metres it is described as one of 'water stress'.[38] Now, 60 per cent of
resources are situated in just nine countries (including Brazil,
Russia, China, Canada, Indonesia and the United States), whereas
eighty countries, representing a total of 40 per cent of the world's
population, are faced with a scarcity of water. The regions most
affected are North Africa and the Middle East, where less than
1,000 cubic metres a year per person (the scarcity threshold) can
be harvested without eating into the region's water capital. Nine
of the fourteen countries of the Middle East suffer from water
scarcity, whereas the possible water harvest is as high as 5,072
cubic metres per person in sub-Saharan Africa and 22,000 cubic
metres in Latin America.[39]

This distribution does not immediately say it all, for the fact
that the United States, Brazil, Russia, South Africa or China are
water-rich countries does not mean that they are free of serious
supply problems. In recent years, for example, shortages have hit
northern China, California, and South Africa – and they affect
social groups and regions within the same country differently,
regardless of its overall water resource level. Thus, in South Africa
600,000 white farmers practising irrigation consume 60 per cent
of the country's water resources, whereas 15 million blacks have
no direct access to water. Factors other than unequal distribution
therefore play a role in the emergence of water problems.

The second category of reasons – waste and, more generally,
inefficient and destructive management of resources – are said to
have caused a total decline of 37 per cent in the possible water
harvest, which now averages 7,400 cubic metres per person. It is
true that the supply uptake has increased hugely over a short
period of time: whereas natural renewal does not raise the level of

water resources, the quantity used grew sixfold from 1900 to 1995 (more than twice the rate of population increase) and twofold since 1975. Agriculture (mainly irrigation) absorbs a world average of 70 per cent of water supplies, rising to 80 to 90 per cent in the under-developed countries; this compares with an average of 20 per cent for industry and 10 per cent for domestic and other uses.[40]

What counts most, however, is the proportion of the water supply that is wasted. Thus, agricultural irrigation systems lose on average 40 per cent of the water they consume; and 50 per cent of the world's treated drinking water is lost through leakage from distribution systems. (In most European countries, both West and East, water pipes date from before the Second World War.) The costs corresponding to these losses have been estimated at 10 billion dollars a year by the Geneva-based UN Economic Commission on Europe.

Everything suggests that, unless there are radical changes, water consumption will continue its high rate of increase as a result of population growth, economic activity and the spread of pollution. The more that the contamination of water on or beneath the surface forces deeper drilling, the more costs will rise and the greater will be the damage to the ecosystem (lowering of the water table). According to the United Nations Industrial Development Organization (UNIDO), industrial activities could absorb twice as much water by the year 2025, and industrial pollution could increase fourfold.

This brings us to **the third category** – the growing number of factors causing pollution. Chief among these are:

- massive use of chemical products and heavy metals (nitrates, lead, mercury, arsenic);

- failure to treat domestic and industrial waste, much of which is directly discharged into rivers;

- massive exploitation of underground water;

- the lack of drainage systems for half the world's population (roughly three billion people);

- land degradation due to deforestation, desertification, etc.;

- floods and other disturbances that are less and less 'natural': the recent overflowing of the Yangtze, which caused 1,500 deaths in one day and threatened a city of 7 million (forcing the local authorities to destroy several dams and to flood large rural and urban areas), says a great deal about human responsibilities and the serious shortcomings of the Chinese authorities.

The fourth major category is population growth. Twenty-five years from now there will be 2 billion extra people on earth, most of them swelling the megacities of megapoverty in Asia, Latin America, Africa, the Middle East and, to a lesser extent, Russia. Some consider that the main reason for the water crisis lies in the growing surfeit of human beings for a limited and shrinking resource, so that less and less will be available for a rising number of people. This is the 'water bomb' that is supposed to be ticking away, according to an argument favoured by the World Bank and its vice-president, Ismaïl Serageldin, who also heads the World Commission on Water for the Twenty-first Century. But it is an argument which overlooks the enormous inequality of consumption between human beings in the North and those in the South. For, as the UNDP Human Development Report for 1998 showed, the richest fifth of the world's population (slightly less than a billion people) account for 86 per cent of the world's consumption. A new-born baby in the West (or a rich one in the South) consumes on average between 40 and 70 times more water than one in the South who has access to water. We should also bear in mind that it takes 400,000 litres of water to make one car, and that most of the 50 million cars produced every year are bought and used by people in the countries of the North.

In the last few years, people have begun to realize the importance of other factors such as economic interests and power play, which are expressed in nationalist geopolitical and geoeconomic strategies with a view to regional hegemony. In general, however, it is still thought that such strategies and conflicts are a result of water scarcity rather than one of its causes. We shall see in the next chapter that this is not a convincing idea. For if stretches of water become the source of disputes and even armed conflicts between states, this is mainly because the states in question are incapable – for political, religious, ethical and economic reasons – of sharing and jointly managing a common resource in the interests of all parties.

The German federal government was right to argue – in March 1998, at the international conference it had organized – that the agreements signed between states bordering the Rhine and the Danube have improved local water supplies. And it is not because water from these two rivers is more plentiful than elsewhere that it has been easier to share this wealth, whereas in other regions local populations have fought with each other over a supposedly scarce resource.

Similar considerations apply to the mode of water distribution within a country, or to the conflict between competing uses considered to be mutually exclusive (for example, between urban and agricultural consumption, between the needs of industrially developed regions and the needs of peripheral underdeveloped regions, or between economic growth objectives and ecological objectives). Here, too, the cause of conflict lies not so much in the fact that Castile has more water than Andalusia, or California less than Nevada or Colorado, as in the fact that socio-economic groups have unequal power to direct and control the modes of resource regulation and distribution.

These other factors lying behind the corporatist and nationalist logics of power – logics that are unsustainable and contrary to solidarity – make it easier to understand how human beings relate to water and to one another because of water. This, too, is why special attention should be paid to the role of an all-conquering Promethean technocratism that has been increasingly called into question.

In short, then, the evidence shows that one of the main causes of the 'water problem' in contemporary societies – at a continental and global as well as a local level – is the political, technocratic, economic, financial, symbolic and cultural power exercised by generations of 'lords' for whom water is itself a source of power, wealth and domination. This is where the main obstacle lies.

CHAPTER 2

THE OBSTACLE: THE LORDS OF WATER

The history of relations between human beings and water – and, even more, of those *among* humans *because* of water – is difficult, tumultuous and fascinating.[41] (It should not be forgotten that in Judaeo-Christian civilization water is associated with the image of an end of humanity that has already taken place once: Noah's flood.) It is a history of sharing and exclusion, cooperation and war, rationality and mystification, art and destruction.

Since earliest times, water has been one of the chief social regulators. The structures of peasant societies and village communities, where living conditions are closely bound up with the soil, have been organized around water. And in the great majority of cases, even when it has been regarded as a common asset, water has become the source of power, both material and immaterial. Only rarely have the members of a community all been on an even footing in terms of water; access to it has nearly always involved inequality.

This points to the crucial importance, both in and for human history, of the efforts that have been and continue to be made to bring about a society where the rights to life are equally accessible to all. It is this pursuit of equality, justice and solidarity that gives meaning to the water revolution and is at the root of the present *Manifesto*.

Relations between communities are also in the great majority of cases marked by varying degrees of conflict. The word 'rival' (or 'rivalry') comes from the Latin *rivus* (stream or brook): a rival is someone who uses the same source of water from the opposite bank – hence the idea of danger or attack.

In short, 'waterlords' have always existed. As in the fable of the wolf and the lamb, the 'lord' lays down the law (the regulatory system) and accuses the weak of wrongdoing that they have not committed. In a system grounded upon the rule of the strongest, such as the one existing today, those who have acquired power (especially economic power) oppose any form of sharing.

The waterlord derives his power from the ownership and control of water, or from the prevailing mechanisms of access, appropriation and use, which enable him to benefit to the maximum from the goods and services that water generates or makes it possible to generate. The waterlord is thus able to increase his capacity for action (in terms of knowledge, information, technology, finance, social relations and cultural power) and to perpetuate his rule.

The legitimacy of his power depends in most cases upon his capacity to provide access (however unequal) to water supplies for the community over which he exerts his authority, by means of systems of harnessing, pumping, channelling, conservation and maintenance. In many situations, the 'lord' draws more legitimacy from the existence of conflicts with other states rather than their absence. It is therefore in his interest that relations with 'rivals' should not be entirely peaceful.

The kind of conflict most frequent today involves competition over uses of water which – as we noted briefly before – become mutually exclusive within a framework marked by lack of solidarity and growing scarcity. One thinks of the alternative uses of water for irrigation or for personal (especially urban) household consumption, or else, as in Spain, of uses that satisfy the needs of people living in the capital and thus reduce the supply available to the inhabitants of Andalusia. The previously mentioned conflict over IBM's pumping of water from ancient deposits in Essonne is another case in point, although there it concerned not so much competing or alternative uses as a choice between use and non-use of certain water resources (exploitation versus consumption/

protection), or rather between use of ancient underground water and use of surface water. The problem was to know who would pay for the increased cost of using surface water (IBM? The purchasers of microchips?).

Similarly, the waterlord has no great interest in the solidarity demonstrated by an equal distribution of the goods and services generated by water, for he believes – and therefore fears – that he would lose part of his wealth. In his eyes, sharing is the origin of risks, since it might allow others to accumulate enough to take away his power. This explains why, in every society, the strongest tend to maintain or create unequal forms of access to basic resources.

Finally, the waterlords derive their power from the symbolic, sacred, mythical aspect of water. In our own high-tech societies, for example, dams have played this symbolic or mythical role; they express man's power to control water and to 'feed' pools and reservoirs with it.

In the 1950s and 1960s, most notably in the Californian desert, the personal swimming pool joined the car as a symbol of freedom in a Hollywood world representative of America triumphant, which thought it deserved such freedom because it had shown itself to be technologically and economically mightier than others. Today, a strong symbolic charge of power and technological progress also attaches to bottled water, perceived as high-quality water that is supposed to be good for health and therefore truly a source of life.

In the light of what has been said above, we may group the 'waterlords' into three main categories: *warlords*, *money lords* and *technology lords*:

- *Warlords* are those whose power and survival continually depend upon violent conflicts or even wars between rivals, either between states or within states over competing uses.

- *Money lords* are those whose power and survival depend upon access to water and upon the rejection of solidarity implicit in unequal distribution of the goods and services generated by water. Today these are mainly forces pushing for the privatization of water regulation systems and the primacy of financial considerations ('shareholder value') over all others in the direction and framing of such systems.

- *Technology lords* are those whose power and survival depend upon faith in the technological imperative (that everything technologically possible should actually be done), and in the notion that human progress stems from social progress, which in turn depends upon economic progress, which in turn depends upon and is determined by technological progress.

Warlords

Conflicts within countries

Conflicts within countries are ubiquitous, both in the Third World and in the rich industrial heartlands. Most often, solutions are found. When a conflict assumes major or critical proportions, it shows that regional or national politics has been incapable of developing and applying an integrated water policy inspired by the primacy of the general interest with regard to a common good or asset (*res publica*) and aimed at fostering solidarity among all the members of a regional or national community. The growing number and intensity of such conflicts are signs of the weakness of the collective regulation system and imply a major fragmentation of relations among the constituted social groups and interests.

The more a society allows the corporatist interests of individuals and groups to become the basis of its own organization and the principle inspiring its operation, the more one would expect there to be a multiplication and intensification of conflicts inside the country – and not only on water-related issues. In such a case, it becomes imperative to draft and approve national water legislation (or simply to respect existing laws where these are inspired by principles of solidarity and sustainability). An interesting example is that of Ecuador, a country which for a number of years – and especially since a land reform amendment in 1994 – has been trying to work out a new law on water. Two contending proposals have been advanced. The first, by the Agricultural Chambers of Commerce, defends the interests of the big farmers and agribusiness and argues that it is thanks to their activities, rather than the peasant economy, that water is used as productively as possible for the country's development. The bottom line here is a call for the privatization of water. On the other hand, the proposal formulated by CONAIE (the Ecuadorian Indigenous Nationalities Confederation) on behalf of the small farmers, maintains that water is a public asset which should mainly serve the equitable development of the country's entire population, and that the food security of the local population should be the number one priority.

In the distinctive world view of Ecuador's native population, man, nature and society are one and the same entity. The main aim of their draft legislation is therefore to ensure that a fair balance is struck between the development of the individual or society and the totality of natural resources.[42] More specifically, this means that water should mainly serve to meet the essential needs of the whole collective, and that the *right* to use water is directly linked to the *duty* to maintain water resources.[43] In theory, this principle is also coming to be accepted around the world as fundamental to the integrated and sustainable management of natural resources: no use without conservation.

Conflicts between countries

Conflicts within countries are, if one may put it like that, circumscribed; those between countries are more serious because of the

forms they may take (up to and including outright war), because of the political, economic and social instability they cause regionally and internationally, and because millions or even hundreds of millions of people thereby perceive millions of other human beings as rivals or enemies.

In the world today there are some fifty local wars between states. This does not mean that artillery and missiles are actually being fired at this moment in fifty different places. It means that, in some fifty parts of the world, neighbouring states are at war with each other for reasons that include water (the river Jordan and the River Senegal regions), or that the guns have fallen quiet but the conflict remains unresolved (the states through which the Tigris and Euphrates pass), or that water is the cause of serious political and economic differences (e.g., the Nile Basin, the Ganges).

The map opposite gives a summary view of a small number of zones of conflict. The examples set out in the following list recall the issue of conflict for the twenty or so that are currently most important.

Examples of inter-state conflicts related to water

Rivers/lakes	Countries involved	Issues
ASIA		
Brahmaputra, Ganges, Farakka	Bangladesh, India, Nepal	Alluvial deposits, dams, floods, irrigation, international quotas
Mekong	Cambodia, Laos, Thailand, Vietnam	Floods, international quotas
Salween	Tibet, China (Yunan), Burma	Alluvial deposits, floods
MIDDLE EAST		
Euphrates, Tigris	Iraq, Syria, Turkey	International quotas, salinity levels

West Bank aquifer, Jordan, Litani, Yarmuk	Israel, Jordan, Lebanon, Syria	Water diversion, international quotas
AFRICA		
Nile	Mainly Egypt, Ethiopia, Sudan	Alluvial deposits, water diversion, floods, irrigation, international quotas
Lake Chad	Nigeria, Chad	Dam
Okavango	Namibia, Angola, Botswana	Water diversion
EUROPE		
Danube	Hungary, Slovakia	Industrial pollution
Elbe	Germany, Czech Republic	Industrial pollution, salinity levels
Meuse, Escaut	Belgium, Netherlands	Industrial pollution
Szamos (Somes)	Hungary, Romania	Water allocation
Tagus	Spain, Portugal	Water allocation
AMERICAS		
St Lawrence Bay	Canada (Quebec), United States	Hydraulic works
Colorado, Rio Grande	United States, Mexico	Chemical pollution, international quotas, salinity levels
Great Lakes	Canada, United States	Pollution
Lauca	Bolivia, Chile	Dams, salinity
Paraná	Argentina, Brazil	Dams, flooding of land
Cenepa	Ecuador, Peru	Water allocation

Note: Of the world's 214 basins mentioned in Corson, 155 are shared by two countries, 36 by three, and 23 by a number up to as many as twelve (including Nile: 9 countries, Congo: 9, Mekong: 6, Amazon: 7, and Zambezi: 8).

Sources: Walter H. Corson (ed.), *The Global Ecology Handbook,* Boston: Beacon Press, 1990, pp. 160–1; and Peter H. Fleick (ed.), *Water in Crisis: a Guide to the World's Freshwater Resources,* New York: Oxford University Press, 1993.

The extensive literature in all languages on this question (mass media included) reflects the wide readership water wars attract.[44]

Causes of inter-state water conflict: why the growing disjuncture between supply and demand is an inadequate explanation

Most analyses of water conflicts highlight factors relating to the growth in water needs combined with the growth in situations of scarcity or limited supply.[45] At first sight it is a convincing argument: the more water supplies dwindle and their quality declines or comes under threat, the more a country's inhabitants will fight with others sharing the same basin to ensure that they are best-off in terms of appropriation and utilization of the available water.

Now, these factors unquestionably do play a significant role. It is no accident, for example, that the sharpest conflicts at the present time – or the most dangerous for the future – are those taking place in the Middle East region, where the availability of water is the lowest in the world.

Nevertheless, the argument based on scarcity tells only half the truth. Other analyses rightly stress the importance of factors such as:

- ethnic rivalry, racism and xenophobia;

- nationalism of every kind;

- struggles for regional political, economic or cultural hegemony.[46]

In the case of the river Jordan, it is evident that the water war is the result and not the cause of conflict between Arab states (especially Syria, Jordan, the Palestinian authority and Lebanon) and Israel. Beyond the historical roots in religious opposition between Jews and Muslims, the war can be traced back to the fact that the Second World War victors satisfied a legitimate claim of the

Jewish people (creation of the state of Israel) but did not satisfy the equally legitimate claims of the Arab, and above all Palestinian, people.

Since then, water has been an exacerbating factor, either creating new hotbeds or fanning long-standing conflicts. Thus, in the Six Day War of June 1967, the immediate occasion was an attempt by the Arab states to divert the waters of the Jordan after Israel had built its 'national waterway', a dispute that flared even though the states in question had signed an allocation agreement in 1964. But here the point is that the Arab–Israeli conflict went far beyond water-related issues. As a specialist on the area has written, 'water is only one aspect of the multidimensional dispute between the Arab states and Israel'.[47] A solution to the region's water problems depends not upon water but upon the political will of the various national leaders to put an end to their decades-long dispute by recognizing the rights of other states to exist and develop.

One can understand why Boutros Boutros-Ghali, the Egyptian former secretary-general of the United Nations, stated as long ago as 1974 that the next world war, if there was one, would be due to water-related conflicts. But it is still an overstatement. For another world war (if one can at all imagine such a horror) would be the result of total irresponsibility on the part of political, economic and religious leaders; in no case could water scarcity be considered a rational 'reason', still less a justification, for its outbreak.

The same considerations apply to the conflict over the Tigris and Euphrates basins, which, for many years now, has poisoned relations among Turkey, Iraq, Syria and Iran. It first became important in the 1960s, when Turkey, as an upstream country (90 per cent of the waters of the Euphrates originate in it), and Syria, made it known that they intended to build a number of dams (13 on the Turkish side) for the purposes of irrigation and hydro-electricity generation – projects which would considerably alter the regional economy and the relationship of forces among the various countries. Tension reached fever pitch in 1974, when Iraq threatened to bomb the Tabga dam in Syria and massed its troops along the frontier, and again in the spring of 1975, when similar threats appear to have been made. In 1987 Turkey offered to export water from the Euphrates to the other countries in the region by jointly building with them a 'peace aqueduct'. But the Arab countries in question rejected the idea, both because of its high cost and because they feared then, and still fear today, that it would give Turkey an unacceptable degree of control over water supplies. Tensions resurfaced in 1990 following completion of the Ataturk dam on the Euphrates. This gives Turkey considerable power over river levels, as it showed by threatening to reduce the flow of water to countries further downstream in order to persuade Syria to withdraw its support for Kurds fighting for their independence in south-east Turkey. At present, despite a number of small advances, the conflict continues to simmer beneath the surface. Turkey has refused to sign the two international conventions to which great efforts were devoted over a long period of time: the Convention on the Use of International Watercourses for Non-Navigational Purposes, and the Convention on the Protection and Use of Transborder Watercourses and International Lakes.

The conflicts between Iraq and Iran (including the war of 1980–4 for control of the Shatt al Arab), Iraq and Syria, Turkey and Iran, Turkey and Iraq, and Turkey and Syria will not cease until the leaders of these countries resign themselves to giving up any aspiration to supremacy or hegemony over the region as a whole. For this is the chief source of conflict in the region: each country still thinks it can gain political hegemony, or at least that it has a duty at all costs to prevent others from exercising too much power. As Jacques Sironneau pointedly remarked, 'the conflict [between Iran and Iraq] to acquire the Shatt al Arab reflects the struggle waged by each of these two countries for regional supremacy'.[48]

In 1995 a mini-war between Ecuador and Peru, centred on the sources of the river Cenepa, resulted in the deaths of more than fifty people. It was not caused by water problems, however, but by the issue of control over an area thought to be very rich in minerals, whose exact shape varies according to who owns the sources of the Cenepa.

As we can see, then – and many more examples could be taken from the list on pp. 41–2 – the warlords are social groups representative of military, economic, ethnic or religious interests which impel a momentum of conflict and domination in order to gain exclusive control over certain resources.

In most of the basins in question, the warlords are located in the upriver states. Asserting a principle of absolute territorial sovereignty, they claim to have exclusive ownership of water resources on their territory (both surface and underground) and the right to use them in any way they see fit. This is naturally rejected by downriver states (Egypt, for example, in relation to Sudan and Ethiopia), which assert the principle of absolute territorial integrity and maintain that such states have the right to benefit from the natural, uninterrupted and undiminished flow of watercourses originating in other countries.

These two principles cannot but lead to conflicts, and indeed they do so. Other principles therefore have been developed and (fortunately) applied. These are: *the principle of limited and integrated territorial sovereignty*, according to which every state has a right to use the waters on its territory, on condition that this does not harm the interests of other states; *the principle of a community of interests*, according to which no state may use the waters on its territory without consulting other states to achieve integrated management based upon cooperation; and *the principle of fair and reasonable use,* according to which each state has a right to use the waters of the respective basin by being awarded ownership and control of a fair and reasonable share of the basin's resources.

It goes without saying that, if all states respected these principles to the letter, the warlords would have much less power to defend. The last two principles, in particular, mark a considerable step forward and have allowed certain conflicts to be resolved.

In the case of conflicts *within* countries, the most significant example is the adoption of *river contracts* in Belgium, Switzerland and France. These consist of an agreement among all the parties concerned (local population, industries, public authorities, the tourist sector, various associations, etc.) for a coordinated long-term management of the river in the common interest. The results have been remarkable in several respects. It should be stressed that the regulatory system is not one-dimensional: it does not involve only states, only public authorities, or only private market actors. Criteria of public utility, economic profitability, social value and environmental sustainability are given equal weight within the system, the aim of which is to arrive at effective solutions that will satisfy all sides.

In the case of conflicts *between* states, the agreement in the American West among California, Arizona and Colorado (see Chapter 1 above) is an application of the principle of fair and reasonable use. The same principle led to the accord between India and Pakistan concerning the waters of the Indus. The conflict was caused by the partition of the subcontinent in 1947 which led to the physical division of the Indus basin between the two countries and to the termination of a long-standing irrigation system, without any clear specification of how the waters should be allocated. As a result, India found itself in the position of being able to control water supplies used for irrigation in Pakistan – a power it used in 1948 to divert them elsewhere. The canals in question were subsequently reopened after energetic protests from Pakistan, and in 1952 talks began on working out a treaty on the issue.

But given the climate of hostility prevailing between the two countries, they did not reach agreement on a unified system that was in their mutual interest. The basin was again divided to India's advantage, while Pakistan was given monetary compensation to build new canals leading from its allocated tributaries of the Indus, and an international consortium agreed to fund a dam that would provide the country with a secure and stable supply of water. Unfortunately, however, the permanent state of tension between India and Pakistan – due mainly to their rival claims on Kashmir (currently divided between them) and to their recent nuclear tests within a few months of each other – has grown worse and even led to armed clashes in August 1998. The water-related disputes have well and truly started up again.

Risks of worsening tension associated with the commodification of water

The example of India and Pakistan again shows that water is not sufficient by itself to explain conflicts, nor to provide a solution to them. It also shows that the principle of *a community of interests* and the principle of *fair and reasonable use* have major limitations. In fact, there is a danger that the principles will not have sufficiently deep roots to withstand pressure from unsustainable logics opposed to solidarity – pressures that will constantly grow over the next 20–25 years owing to population growth and the intensification of technological and financial rivalry among the most developed and aggressive powers to secure world markets and continental or global economic dominance.

Ten years from now, India will have an extra 250 million people, Pakistan's population will have soared from 128 million (in 1994) to 210 million, Turkey's will be approaching 80 million, Ethiopia's 85 million, Egypt's 82 million – and we could go on. Furthermore, the logic of competition, the primacy of profits and the race to create global giants with huge financial and industrial power will add further venom to economic and geopolitical struggles between countries to control natural resources. If these are to be the main trends, what value can the principle of common interests have? How much resistance can there be from the principle of fair and reasonable use? Will not both principles simply vaporize at the first difficulties and allow the warlords to continue exercising their power without obstruction? Under such conditions, it is highly likely that water wars will increase in number and intensity.

Nor does it appear that the might of the money lords and technology lords is about to weaken. We hear it said more and more often that the solution to water-related conflicts must involve economics, and even that the market must be allowed to establish a match between needs and supply. In this context, a unique role is supposed to fall to technology, technological-commercial innovation.

The journalist Jean-Paul Besset identified this feature of the *zeitgeist* when he entitled an article in *Le Monde*: 'Water: War or the Market'.[49] But it is a pity that, instead of taking the two terms as non-solutions, he opted for the market as the inevitable solution to the water problem, thus siding with the money lords.

Money lords

The thesis which is slowly but surely spreading among ruling groups in the developed countries, and also gaining influence among the public in general, rests upon a few apparently simple and true ideas whose scientific relevance and empirical validity are not as evident as we are led to believe. What are these ideas?

- The enormous waste in the use and management of water – a phenomenon discussed in the previous chapter – is supposedly due to the fact that most of our societies have hitherto considered water to be a social asset rather than a commodity.

This has artificially kept the price of water very low and encouraged careless, wasteful and inefficient use, most notably in agriculture (which, as we said before, accounts for more than 70 per cent of world water consumption) and in the home. Water should therefore no longer be treated as if it were available in abundance, says the World Bank's Ismaïl Serageldin, but should be redefined as an economic asset. Or, in the words of Mark Rosegrant, a researcher at the highly influential International Food Policy Research Institute in the United States: 'We must get away from the notion that water is a free good'.

• We are no longer in the age of abundance, this argument goes; the pre-scarcity era is over; water availability is at crisis levels (the 'water crisis') and our societies are in danger of sinking into a state of general shortage. Too many people are seeking to obtain water, at a time when the amount available per person is dwindling each year. This is the main reason for water wars, and to avoid them in the future it will be necessary to raise the price of water to reflect the new age of scarcity. Efficient management of water requires real prices set by the market, since the more expensive water becomes, the less of it is used. The conversion of water into a commodity cashable on international markets, or even on a global market, is thus supposed to be the best guarantee against water wars. Water, once it is recognized as an economic asset, will be a factor for peace; market prices will bring peace in their train.

• The market, it is argued, will lead to efficient distribution and use of water. So long as certain precise criteria are respected (such as a clear definition of markets and assured property rights), markets will give individuals and countries a greater opportunity and capacity to develop, transfer and use water resources in a manner beneficial to the whole world.

- The champions of the *zeitgeist*, while expressing a wish to move beyond ideological disputes, maintain that privatized regimes of ownership, of infrastructural construction and management, of distribution and purification, offer a more efficient and profitable solution, and one more in the general interest, than regimes of public, national, state or municipal ownership. The private sector is a symbol of efficiency, profitability, flexibility and equity, whereas the state (both central and local) is synonymous with bureaucracy, inefficiency, rigidity, lethargy and corporatism. In places where the public authorities have shown themselves manifestly incapable of providing an assured supply of water, even in high-need areas such as Manila, Buenos Aires, Hanoi, Mexico City or Gabon, privatization would supposedly allow access by the greatest possible number of people. 'To offer private enterprise the opportunity to run an efficient system appears today to be the means of providing the best services to the poor at the most favourable cost', Serageldin concluded in 1995.

As we can see, these are 'strong' ideas which have the advantage of appearing to be clear, simple and pragmatic. They inspired the water privatization under the Thatcher government in Britain in 1989, and they mean that the Labour government which returned to power in 1997 has no intention of going back on that decision. Renationalization or resocialization of water has no place on Tony Blair's agenda.

The ideas presented above raise a number of major problems. Let us now look more closely at each one in turn.

The main factors responsible for water shortages and inefficient management

Are shortages and scarcity due to the fact that water has not been considered an economic asset? This argument, which contains no more than a grain of truth, holds the artificially low price of water mainly responsible for the huge waste of the past fifty years in the use and management of water. In reality, however, not only has the price of water shot up everywhere in the last ten years without reducing waste (in many of the world's cities the water bill eats up 8 to 9 per cent of average household income), but there is plenty of evidence that the factors mainly responsible for waste and general inefficiency have been and still are:

- agricultural overexploitation;

- industrial pollution;

- the lack of a long-term vision involving integrated global planning and management, or the inability to apply it effectively and coherently because of the economic and financial interests in play.

Three examples taken from India, but replicated in hundreds of situations in many different countries, will serve to illustrate this point. The first example concerns the disastrous state of underground water resources in several parts of the country. Irrigation and rural needs account for 90 per cent of India's aggregate water consumption. But since the 1960s intensive agriculture has driven the extraction of underground water to ever higher levels and, through the use of fertilizers and pesticides, caused massive pollution over very large areas. On top of this has come the rapid growth of cities. The case of the 'thirsty' city of Jodhpur, whose water system literally exploded, can be found in Indian school textbooks. The agricultural revolution has transformed water needs in regions, even desert regions, where they used to be more easily satisfied. Forced to compete to survive, peasants have abandoned low water-intensive crops for high value-added crops that use very large amounts of water. The low price of water has had only a marginal influence: the deep changes in Indian agriculture have followed macrohistoric and political-economic trends which go well beyond the country's particular situation.[50]

The second example concerns the almost irreversible degradation of the river Yamuna, the main source of water for the city of Delhi, which has been reduced to a receptacle for all kinds of waste material. The pollution and general unwholesomeness of its water have passed all critical thresholds. According to Indian experts in the field, the Yamuna catastrophe is due mainly to Delhi's chaotic development, and especially to the evolution of its road system and industry over the past forty years, despite the numerous plans and the large sums of money invested to counter or repair the river's decline. The Yamuna case is one of the clearest examples of mismanagement of water in an urban environment, the responsibility for which lies with the city's leaders in general.[51]

The third and final example concerns another river, the mineral-rich Damodar, which our source describes as probably the worst polluted in India today, because of the heavy industry implanted all along its banks.[52] Even if the price of water had been much higher, it would have had no significant effect upon the development of such industry on the Damodar. Possibly – and then only in the last few years – it would have encouraged a quest for less polluting production processes, but neither the technological knowledge nor the awareness of industrialists, politicians and the public at large worked in favour of such innovations.[53]

It might be thought that these three examples offer a demonstration based upon an anecdotal approach. To avoid this, let us now consider the phenomenon of *water pollution* more generally. This very trend is undoubtedly one of the main roots of the water crisis that has increasingly affected both the developed countries and the Third World since the early 1960s. In the richest and most developed countries, huge anti-pollution programmes have slowed (sometimes quite significantly) the rate of contamination of continental waters caused by domestic and industrial waste carrying organic fermentable matter. But this has been replaced by chemical pollution, especially toxic metals and organic compounds with high levels of acute or chronic toxicity.

The latter kind of pollution, coming from urban and industrial discharge spread over huge areas, affects not only surface water but increasingly also underground sources. At present, surface water pollution has reached considerable levels everywhere in the world. In France, for example, industrial and urban waste was running in the mid-1990s at approximately 8,800 tonnes a day of oxydizable matter. In North America, the situation of the Great Lakes is particularly catastrophic and their pollution is now considered irreversible in the cases of Lake Ontario and Lake Erie. The Environmental Protection Agency estimated in the early 1980s that, out of roughly a hundred thousand sites discharging industrial waste with dangerous chemical substances, more than two thousand were directly contaminating underground waters.[54] Application of the 'polluter pays' principle – whereby agricultural and industrial users (which together take 90 per cent of water in the developed countries and 95 to 96 per cent in the poorer countries) pay an increased price for water – has only slightly inflected the advance of pollution. Worse, since the mid-1980s the passing on of costs to users has led polluters in the developed countries to export or relocate their polluting activities and products towards the poorer countries, where the costs to be paid are really so very low!

Water is more than an 'economic asset' and should not go the way of oil
The idea that water should be considered mainly as an economic asset or a cashable resource, and that market laws will therefore allow the problems of scarcity or even inter-state conflict to be solved, is highly simplistic. It is based upon a purely ideological choice that highlights only one of the many dimensions peculiar to water, emphasizing economic value to the detriment of all other values. This ideological choice is in turn based upon an

assertion that the market is the principal mechanism, superior to all others (political regulation, cooperation or solidarity) when it comes to the optimum allocation of material and immaterial resources and the most efficient distribution of the wealth produced. One can share or reject such an ideological choice. But even its own champions cannot deny (unless they are completely blind to reality) that there are no other sources of life (except air) comparable to water within the ecosystem Earth; that it is a unique resource, quite unlike other resources, to which human beings must have recourse in order to satisfy their basic individual and collective needs. Its unique nature depends, among other things, on the fact that nothing can replace it. Oil can substitute for coal, and nuclear energy for oil; one can substitute rice for wheat or take the train instead of an aeroplane. One can practise a money economy as well as a non-money or non-price economy (where things are free). But it is not possible to replace water and go on living. Yet one of the distinctive principles for properly functioning market mechanisms is that it should be possible to replace certain goods (factors of production or products/services) with other goods – hence the function of relative prices in reflecting the comparative use value of such interchangeable goods and services.

It is thus an essential feature of the market that one should be able to choose among several goods of the same or a different nature, using for that choice such criteria as price and quality. This is what constitutes consumer or producer freedom. To have access to water, however, *is not a matter of choice*. Everyone needs it. The very fact that it cannot be replaced with anything else makes water a *basic asset* that cannot be subordinated to a single sectoral principle of regulation, legitimation and valorization; it comes under the principles of the functioning of society as a whole. This is precisely what is called a *social asset*, a *common good* basic to any *human community*.

In recent years, the integral character of water as a natural resource has been commonly expressed in the principle of 'natural hydrological unity', which underlies the conception and implementation of an integrated water policy.

Now, the fact that a social asset is limited in quantity does not mean that its integrated management by society as a whole should take account of nothing other than the value parameters of a capitalist market economy, which imply the private appropriation, ownership or management of this non-substitutable common social asset. The transformation of water into a commodity along the lines of oil is an aberration bound up with the economism currently dominant among the classes in power – a way of thinking which reduces everything to a commodity and every value to market exchange value.

It is one of the beliefs firmly held among business circles that money is the chief measure of all things. But no less important, in the same order of ideas, are the fantasies that economists entertain about the 'law of supply and demand', and their conviction that water does not escape the 'natural' law of optimization represented and measured by the point at which the two intersect. In the end, this conception rests upon woolly and ambiguous inferences from the shortage and growing scarcity of water. It seems to consider this shortage and growing scarcity as an irreversible fact that can, at the most, be managed or moderated but not fundamentally altered. More specifically, the idea is not that the operations of price and market should enable every human being and every human community to have access to water, but that they should allow *efficient management of shortage or scarcity*, without 'dipping into' the water capital available as a result of the natural renewal of resources. The conversion of water into an economic asset is not supposed to lead to its accessibility for all of the world's population, but rather to what it is claimed to be an 'economically rational' management of a limited resource whose accessibility should be regulated by the solvency of users competing with one another.

Responsibility for access, costs and quality is not a purely individual matter but implies individual and collective rights and duties

This brings us back to the fundamental point that access to water is not a question of choice. Choice intervenes only in relation to the forms of access or use: surface water versus underground water, appropriation and distribution of water supplies, water for agriculture, industry or household needs, tap water or bottled water, and so on. Access to water for the satisfaction of the vital needs of every human person and community is an objective of paramount importance, a *basic obligation* for human society *whatever the cost*. Of course, common sense tells us that the ingenuity, wisdom and skill of human societies will lead them to reduce costs by as much as possible, especially human, social, economic and environmental costs.

The consumption and conservation/protection of water evidently involve many different costs: human, economic, social, political, individual and collective. These costs are neither mutually exclusive (on the contrary, they are interactive and interdependent) nor mutually substitutable (one category of costs cannot be replaced by another). The effectiveness of an integrated and sustainable management of water comes from a cumulative overall counting of all these costs as a whole. In practice, this means that *every society must collectively cover all the costs* of collection, purification, storage, distribution, utilization and recycling necessary to guarantee *basic access to water for all*.

Basic access to water for every human being should be understood as referring to the quantity and quality of water that he or she requires to live as an individual (and family). Similarly, basic access to water for every human community refers to the quantity and quality of water necessary to satisfy collective needs and to ensure the basic economic and social well-being of all its members.

Still more than in other fields (minerals or energy resources), basic access to water is *a fundamental political, economic and social right for both individuals and collectives*, since the biological, economic and social security of every human being and every human community depends upon enjoyment of that right.

Such a right cannot therefore be bought and sold, quoted on the stock exchange, traded or swapped,[55] any more than modern democratic constitutions allow people to buy and sell each other's bodies. Recent trends, however, such as the authorization of patents for plants, animals and human genes – in the United States in 1995 and the European Union on 12 May 1998 – are opening the way to *privatization of the human* and *commodification of the living*.

Human health is closely bound up with secure basic access to water, as we can see from the fact that problems of water quantity or quality are at the root of 85 per cent of human illness in the poor countries. Health protection is thus much more than a matter for personal action by individuals; it is a fundamental task for the collective as a whole. One cannot ask each individual to assume responsibility via price mechanisms for the costs of planning, building and maintaining infrastructure and for improvements to the water supply in both town and country. These are all social, collective costs, which cover the whole range of negative and positive externalities involved in the production, distribution and use of any basic resource. It was not without reason that, after the early privatization of water production, distribution and use in nineteenth-century European and American cities (at the time of rapid industrialization and triumphant national–urban capitalism), the private companies in question transferred ownership and responsibility to the state. For the private sector and the market cannot control that which is, by definition, external to them. And in the case of water, the externalities are sizeable and crucially important, not only locally but at the level of world society and the ecosystem Earth.

Responsibility for the equation: access = security = health, has to be collective, as also must be the awareness and expression of it within each human community and in terms of the security of the planet. Democratic exercise of this responsibility is part of the culture of social coexistence: it fosters a sense of belonging to the same community and helps to feed and strengthen the practices of sustainable management of water resources based upon solidarity.

In sum, the argument that users should be made to pay directly the costs of efficient water management presents a number of major weaknesses. In the United States, where this principle has been applied to health in such a way that the market has become the main form of regulation of commodified goods and services, some 41 million people out of a total population of 280 million were reported in the press in July 1998 to have been without medical cover in 1997. And this is in a country that is the world's leading economic, financial and military 'power'!

What, then, should be the basis for providing and sharing the costs and benefits of the access–security–health nexus? What should be the criteria of selection and evaluation?

As we shall see in the next chapter, it is by treating the world's water as a common human heritage structured around local communities that solutions must be found for an integrated, solidaristic, sustainable and efficient management of water, including the determination and assumption of the relevant costs.

Why the commodification and privatized management and ownership of water resources are in favour

How are we to explain that, despite all the above points, the pressure for water commodification and for the privatization of regulatory systems has gained so much ground, especially (and it would seem paradoxically) in the countries of Latin America, Africa, Asia and the former 'socialist' bloc around the USSR? The answer is not easy. But this is where the role of the 'money lords' is most apparent.

Roughly speaking, three sets of reasons may be identified. *First of all*, the pressure for water commodification and privatization is not an isolated phenomenon. In fact, it is a late expression of a general trend which for at least thirty years has affected all other fields of economic life in developed societies, stemming most particularly from the United States. No public sector, no public service and no public asset has escaped: whether it is a question of mail services, telecommunications, gas and electricity, urban transport, railways, airlines, health, education and training, social security, or even national statistical offices, all have been privatized in whole or in part, with the precise form varying from country to country.[56] This phenomenon is in turn part of a greater ideological, political, economic, social and cultural transformation of regulatory systems, and a turnaround in the relationship of forces among constituted social groups. Since the 1970s, this sea change has enabled forces linked to financial, industrial and commodity capital to capture powers of decision, direction and control with regard to the allocation of material and immaterial resources and the distribution of productivity gains, at local, national, continental and global levels.[57]

No one can deny that the forces and logics of the capitalist market have won power in this way. Directly or via the state, it is these forces, especially those linked to finance capitalism, which more and more govern the principles and forms of the distribution of the world's wealth and, by this means, the forms and priorities of the needs to be satisfied.[58] Thus privatization, deregulation and liberalization have become the watchwords of the regulatory systems of world governance in the areas of economics and relations between economic and political actors. In many countries, the state and politicians have given up their role as chief locus of regulation and handed it over to globalized private corporations and the financial markets. The former US Treasury Secretary, Roger C. Altman, has summarized the present situation well in the title of an article: 'No country is beyond the financial market's power'.[59]

The reasons why water took longer than other goods or services to follow this trend have to do with the peculiar irrationality and unjustifiability of its commodification and privatization, as we have seen above. But the dams have burst, so to speak, in the last few years, because of the third set of factors that we shall examine shortly.

The *second set* mainly concern the impact on the underdeveloped countries of recent changes in international aid and cooperation policies. Following decolonization (more or less complete by the end of the 1960s), the attitudes of developed countries toward their former colonies were either paternalist or neo-colonialist; it was rare for the countries of Western Europe and the United States to adopt clear and sincere policies of solidarity and cooperation. If one excepts some remarkable principles that inspired the definition and application of the Yaoundé and Lomé conventions until a few years ago, it must be recognized that the proclamation of a New World Economic Order in 1974 (following the oil crisis), a World Development Decade (from 1977 to 1987) and then an International Drinking Water and Sanitation Decade (1981 to 1990) were far from producing the anticipated results. Aid remained essentially tied to the interests of donor countries. The terms of trade continued to evolve to the advantage of developed countries, and at the end of the Decades Third World countries found themselves up to their ears in debt. Instead of cooperation and solidarity, they were handed down restructuring programmes by multilateral bodies skilled either in financial engineering (the IMF and the World Bank) or in commercial engineering (GATT and, since 1994, the World Trade Organization).

In the course of the past twenty years, the action programmes of intergovernmental political institutions which, for better or worse, promised some kind of cooperation and international solidarity, have been increasingly marginalized. It is no longer UNESCO, WHO, the Food and Agriculture Organization (FAO) or the International Labour Organization (ILO) which guide and manage the search for solutions to the world's problems; such tasks now fall to the international financial agencies, the World Bank and the IMF. According to Article 1 of the World Bank's statutes, however, its purpose as an institution is the promotion of private investment abroad. Whether a loan will boost development of the private sector and, more generally, private production of goods and services traditionally operated by the state or cooperatives is one of the IMF's main criteria in arriving at its credit decisions.

The main development-funding institutions linked to the multilateral agencies of the United Nations and the 'holy trinity' of IMF/World Bank/WTO now make privatization a compulsory step in what they call the modernization of the water services management sector (major improvements, irrigation, collection and distribution of drinking water, urban sanitation).

This political choice is not unimportant, at a time when national states are withdrawing from governance of the international economy on the grounds that the global economy supposedly requires more flexible, less formal, more networked systems of governance – characteristics which multilateral agencies such as the IMF, World Bank and WTO can offer more effectively than states are able to do.[60]

One of the basic reference texts guiding the activity of the World Bank and the IMF is Gabriel Roth's work on public services in developing countries.[61] His main argument (after showing why piped drinking water has not so far greatly interested the private sector)[62] is that the main obstacle to privatized distribution of drinking water has been a political refusal on the part of governments. Roth's favourite model is French-style privatization.

His proposals tend to make an even clearer separation than in the French case between the national government (which would provide the funding with help from the World Bank and IMF) and the private sector (which would ensure proper functioning at market prices).

Roth's work was published in 1987. The meagre results of the Development Decade and the Water Decade, the final crisis of the USSR and 'actually existing socialism', the transition of all the socialist bloc countries (including China) to a market economy, the structural weaknesses of the newly independent African and Asian states: all this gave an even stronger impetus to the privatization and marketization of the functions of direction, control and management of the economy of underdeveloped countries and their relations with the developed countries. 'Forget about aid, think business!' has become the prescription of the World Bank, IMF and WTO, backed up by the governments of the developed countries.

The World Bank does recognize, however, that local water markets should not be expected to lead to better distribution of water among the various sectors of use, or to improvements in the quality of water (in the foreseeable future, at least). 'The public authorities should therefore take *ultimate* responsibility for the distribution and conservation of water. This requires an investment and allocation policy (supply management) and also, in order to influence consumer behaviour, measures that involve direct regulation, technological innovation, financial stimulation and appeals for self-control (demand management).'[63]

Notwithstanding this, the World Bank considers that the state's role should be to set the game rules (including water prices, though at the marginal cost of production) and to promote market mechanisms, and no longer to involve itself directly in water management. Management and/or ownership should be left entirely in the hands of the private sector. But the state should also provide the security of the law for water ownership and rights transfers, as well as defining and enforcing quality standards for safe drinking water.

The *third and final set of reasons* for the increased commodification and privatization pressures of recent years were clearly expressed in a short article by *Financial Times* journalist John Barham, 'How to Sell the World's Water Industry', in which he reported on an international water industry conference held in Istanbul in late September/early October 1997.[64] Water, he remarked, is 'the last frontier in privatization around the world'. It is now attracting the attention of the private sector because 'although it is often regarded as so basic to human life that it cannot be treated like any other commodity, it is becoming so scarce that it now commands high prices'. 'Even the poor' – here he reports the views of a director at the International Finance Corporation, the group responsible for private sector projects of the World Bank – 'are prepared to pay very high prices for water.' In these conditions, the water sector has become a potentially lucrative market. There is a certain reticence in financial circles, however, where some believe it is still too risky politically, economically and socially. In other words, they realize that they would not be able to make profits regardless of the human and social costs – that is, without guaranteeing access to water even for those unable to pay. But John Barham has some words of reassurance for them: the only functions of the state are to carry out the necessary economic reforms (of which the most fundamental is 'putting a market price on water'), to guarantee fiscal stability and to give financial support for major infrastructural works. The number of water privatizations is still quite modest in comparison with telecommunications, energy or other formerly public sectors. Nevertheless – he concludes – by turning water into an economic asset, a cashable resource, governments are making the sector as attractive as the others are to the market.

This simple statement of the case is an accurate reflection of the feelings, views and strategies of industrial and financial circles. For the money lords, it is quite clear that:

- water will become more expensive;

- even the poor will be prepared (compelled) to buy high-priced water, not only in bottles but from the tap;

- the bottled water industry will continue its remarkable advance of the last ten years as diversification opens up huge new markets;

- both rich and poor countries will have to invest huge sums in the construction or renovation of water production, distribution and treatment infrastructure;

- the needs that will become most pressing over the next twenty years will be in the 650 world cities with populations of over a million: more than 600 of them will be located in Africa, Latin America, Asia and Russia;

- whereas state regulation and controls on private intervention used to mean that the distribution of drinking water was the only really profitable sub-sector, the situation has changed considerably over the past ten to fifteen years, so that the profitability of new investment, even in sub-sectors such as water purification, has become exceptionally high and secure for private capital.

The money lords have managed to convince most of the world's politicians that these points are correct and important. Thus, according to the fourth principle of the Dublin Declaration, approved at one of the most significant intergovernmental conferences of the 1990s on water issues:

Water has an economic value in all its competing uses and should be recognized as an economic asset. *Following this principle* [emphasis added], it is especially crucial to recognize the basic right of all human beings to have access to drinking water and sanitation at affordable price. Past failure to recognize the economic value of water led to wastage and to uses that were harmful to the environment. To manage water as an economic asset is an important path to the achievement of efficient and equitable use, and to the encouragement of the conservation and protection of water resources.

The other three principles of the Dublin Declaration are:

- fresh water is a limited and vulnerable resource, essential to life, development and the environment;

- the development and management of water should be based upon an approach that involves the participation of users, planners and political decision makers at every level;

- women play a central role in the provision, management and safeguarding of water.

With these principles under their belts, the money lords can get seriously down to work!

The water industry: it is time to get to know it better

We do not have sufficient knowledge of the *water industry*. Whereas the telecommunications, energy, health and transport industries (still partly public or subject to state regulation) figure in national and international statistics, strategic economic analyses and international comparisons, nothing comparable can be said with regard to water. The Statistical Office of the European Union compiles a lengthy section on telecommunications but only two to three pages on water, and in the EU's annual *Survey of European Industry* – which contains a vast amount of information – the section on water cuts a Cinderella-like figure with no more than five pages of fleeting data and analysis. We ourselves have found it impossible to construct a reliable table of the main public and private water corporations operating in the Triad countries (North America. Western Europe, Japan), in accordance with their turnover, income, employment, profits and evolution over the past ten to fifteen years. Nor, for lack of coherent sources and sufficient time, have we been able to establish a list of all the cities in the world with populations over 250,000 which have privatized their water systems.

We have been able to consult, however,[65] a number of articles, working notes and documents – including a *Global Water Industry Outlook: Industry Report*[66] containing financial data especially for the ten major US private corporations in the field, and a *Survey of the European Water Industry* produced by the European Public Service Confederation (London, 14 November 1994) at the request of the Directorate-General for Social Affairs of the European Commission (not, be it noted, the Directorate-General for Industry).

None of the major conferences of recent years listed above (pp. 24–5) have paid special attention to the water industry, the water corporations or their markets and strategies. But information and analysis abound on the state of water regimes in different countries and on public/private water management partnerships.[67]

Let us express a wish here that this poverty of knowledge will be eliminated as quickly as possible – for the benefit not only of researchers, but, above all, of political leaders and public opinion in general. Maybe it is something that might be done by the International Water Office in Limoges (whose information and documentation service EAUDOC mentions 'those active in the water sector' as one of the topics it handles), or else by the International Water Secretariat, the Global Water Partnership or the World Water Council. The more that private capital establishes itself in the water sector, the more it is in the interests of all parties concerned to have rigorous, accessible and detailed knowledge of the industry.

Embryonic globalization

For the present, things are still pretty 'homespun'. Everyone knows, for example, that French corporations – in particular, the two media-nominated 'water giants', Générale des eaux (of the Vivendi group) and Suez–Lyonnaise des eaux – are far and away the world's largest water distribution companies. Vivendi is the leading world water operator (with a turnover of $7.1 billion in 1997) and is also involved in collective services in the environmental sector, energy, urban sanitation and public transport. Lyonnaise des eaux (now part of the Suez group) is second to Vivendi with a turnover of $5.1 billion in 1996, but it is number one internationally ($2.9 billion in 1997, against $2.2 billion for Vivendi) and its subsidiary, Degremont, is the world leader in water treatment engineering.

In 1997 the French no. 3 and no. 4 – Saur and Cise – merged with each other within the framework of the expanding Bouygues group (the world number one in civil construction). Saur–Bouygues supplies water to 34 million people around the world, compared with 70 million each in the cases of Suez–Lyonnaise and Vivendi.

Following the privatization of water in the United Kingdom in 1989, the British companies in the field (initially eight but reduced to five through takeovers and mergers) have begun to take an interest in international markets, especially Severn–Trent and Thames Water. They are still a long way behind the French corporations, however, as are the Americans (such as Bechtel), the Germans and the Japanese. The world presence of Lyonnaise des eaux is for the moment unique (see the table on the next page), but in 1998 and 1999 Vivendi gained a lot of foreign markets and is now not far behind.

Since the measures to open up public services and to deregulate national public monopolies, German companies such as RWE have shown great dynamism. In 2000 RWE bought Thames

The globalization of a water company: the case of La Lyonnaise

Company	Country	% of capital held	Sector of activity
Aguas Argentinas	Argentina	25.5%	Water
Lyonnaise–Australie	Australia	100.0%	Water
Sita	Belgium	100.0%	Waste management
Aquinter	Belgium	45.0%	Water
Sofege	Belgium	100.0%	Water
SS2	Czech Republic	51.0%	Construction
SMP	Czech Republic	51.0%	Construction
Lyonnaise (C2)	Czech Republic	100.0%	Water
Lyonnaise Chine	China	100.0%	Water
Eurowasser	Germany	49.0%	Water
Brodrier	Germany	25.0%	Construction
Aguas de Barcelona	Spain	23.0%	Water
Cespa	Spain	100.0%	Waste management
Lyonnaise Pacific	French overseas depts	100.0%	Water
CEM	Hong Kong	20.0%	Energy
SAAM	Hong Kong	43.0%	Water
Lyonnaise Indonésie	Indonesia	100.0%	Water
Crea	Italy	49.0%	Water
Sita	Italy	100.0%	Waste management
Lyonnaise–Lituanie	Lithuania	100.0%	Water
Lyonnaise–Hongrie	Hungary	100.0%	Water
Lyonnaise–Malaisie	Malaysia	100.0%	Water
Safege Roumanie	Romania	100.0%	Water
Sita Clean	UK	100.0%	Waste management
Essex & Suffolk	UK	99.0%	Water
Lyonnaise UK	UK	80.0%	Water
North-East Water	UK	99.0%	Water
General Water Works	UK	26.0%	Water

Source: European Water Industry, *EPC Survey,* EPSC, Brussels (n.d., probably 1994)

Water, world no. 3 in the water industry. Its aim, as publicly stated, is to become one of the largest *multi-utilities* corporations in the world. Financiers already rejoice at the prospect of huge profits that will follow major battles among the most important European water companies for shares of the single European market, now that it is deregulated and open to international competition.

As more cities privatize water distribution, the opportunities for profit increase; hence the growing interest of private capital in taking charge of the drinking water sub-sector. To take only the example of Lyonnaise des eaux, the number of large cities which have entrusted it with management of water and/or the environment has been growing at a fine pace. The following fourteen made the move in 1997 and 1998 alone:

Manila	Philippines	Water and sanitation
Budapest	Hungary	Water management and distribution
Cordoba	Argentina	Distribution of drinking water
Casablanca	Morocco	Electricity and water distribution and sanitation
Jakarta	Indonesia	Water production and distribution
Medan	Indonesia	Water production
La Paz and suburbs	Bolivia	Water and sanitation
El Alto	Bolivia	Water and sanitation
Indianapolis	USA	Sanitation
Milwaukee	USA	Sanitation
Tianjin	China	Water production
Zhongshan	China	Drinking water
Ho Chi Minh City	Vietnam	Water production
Potsdam	Germany	Water and sanitation

Source: BOUTONMAR.CO Htm/ResAnchor www.suez-lyonnaise-eaux.fr

The more the private sector gets involved in water, the more comfortable and reassured it feels about intervening in the sanitation sub-sector. The phenomenon is already frequent and significant, and if it were to assume vaster proportions the whole system of access–safety–health would be more and more subordinated to the logic of the market and financial profitability: the most decisive goal for each company would ultimately be to increase the value of its shareholders' capital.

The private Swiss bank Pictet announced the launch in January 2000 of a unit trust geared to the stock market values of eighty water companies (between $50 billion and $100 billion). This is the first time in financial history that such a public fund has been exclusively devoted to water securities. The companies comprising its reference portfolio were chosen by Pictet because of their high profitability and great hopes for their long-term appreciation. If this initiative proves convincing, we shall see the birth of a global water market dominated by the logic of finance. Companies will be driven to keep raising their return on investment, lest they see capital desert them for more profitable enterprises.

Limits and problems of water privatization

Why should the growing hold of industrial and financial interests in the ownership and management of water be a source of concern? Is this not the sign of a pointless and mystifying hostility to the private sector, when we consider that European Union data for the twelve member states in 1992 showed that the weight of private interests was still relatively low – an average of 16 per cent weighted by population for direct private management/ownership, and 20.5 per cent for delegated private or mixed management/ownership (see the table on page 72)? The concerns are justified, not only because the situation has changed rapidly since 1992, but also because of the two most important country experiences of water privatization thus far: namely, in the United Kingdom and France.

What is so interesting about these two cases of privatization is that they differ both in their ground rules and in their operational forms.[68] The French system is based on the principle of delegated management of a public service to private companies. According

Abstraction, purification and distribution of water: forms of management in the 12 EU member states, 1992

Percentage of population served by type of management and by member state, 1992

	FORM OF MANAGEMENT				
	Direct public	*Direct supra-municipal*	*Delegated public or mixed*	*Delegated private or mixed*	*Direct private*
12 EU★	37	11	16	21	18
Belgium	5	0	90	5	0
Denmark	67	0	33	0	0
France	23	0	2	75	0
Germany	35	20	30	15	0
Ireland	100	0	0	0	0
Italy	72	23	1	4	0
Luxemburg	100	0	0	0	0
Netherlands	15	0	85	0	0
Portugal	92	0	8	0	0
Spain	48	11	12	29	0
United Kingdom★★	3	9	0	0	88

★ Average weighted by population, excluding Greece.

★★ England and Wales: direct private management; Scotland: direct regional
management; Northern Ireland: direct public management.

Source: EUREAU.

to the water law of 1992, the state sets the general rules while the managerial oversight is the responsibility of the communes and their various bodies. The division of the country into six major water basins created the framework within which its water resources are supposed to be managed in a balanced way. Local authorities may either entrust the management of water services to a private company or take direct charge of it themselves. In France there is no real scarcity of drinking water. Problems exist with regard to its purity and the fight against pollution and waste.

Water is distributed through a system of delegated management in the case of some 85 per cent of the population, while private companies already account for more than 35 per cent of sanitation services and are tending rapidly to increase this proportion, according to a parliamentary report drafted by Ambroise Guelle, a member of the French National Assembly.[69] As we have seen, three large private corporations – Vivendi, Lyonnaise des eaux and Saur– Bouygues – share virtually all the water management in a monopolistic position at local authority level. Another private French company, Danone, is the world number two in bottled water. It is thanks to their decades of experience in water services that Lyonnaise des eaux and Vivendi have managed to expand so widely abroad and to become the world's two largest water companies. This is a strategically important matter for the French economy, and it is hard to see a government opting for any change of direction in water policy and management that might weaken or endanger the dominant position of French water companies on world markets. Despite the positive aspects, however, water management in France ('part of the common national heritage', according to Article 1 of the 1992 law) is increasingly geared mainly to the benefit of economic and financial interests. Nor is it exempt from a hefty dose of corporatist techno-bureaucratism.

First of all, the price of water has constantly increased in recent years (Ambroise Guelle speaks of a price explosion), and this has enabled the private companies to raise their profit levels quite significantly since the middle of the 1980s. Between 1990 and 1994, the average rise was 50 per cent and in some towns (Grenoble, for example) the price tripled; in Paris the increase was 154 per cent. Since privatization started to spread rapidly in 1994, the water industry has become a sector with an especially high rate of return on capital. In the 1990s the profits of Lyonnaise des eaux from its water distribution activities made up 60 per cent of its total profits, although water distribution represents only 25 to 30 per cent of its annual turnover. If the price of water in communes where water services are delegated to private companies is compared with the price where the commune directly provides such services, it emerges from Guelle's parliamentary report that the most expensive of the 40 towns with a population above 10,000 are those with a privatized distribution system (this is true of 17 of the 20 most expensive towns, whereas 13 of the 20 least expensive have a system of local authority management).

The second major defect in the French system, according to the same parliamentary report, concerns the lack of transparency in the awarding of management concessions and the numerous corruption scandals over the years. The logic of profit making at the expense of the state and the population is at the root of what is questionable in the French model. Indeed, the French experience should give cause for thought about the ethical aberrations of a politics and economics that allow private subjects to make a profit out of a common, vital and non-substitutable heritage of society as a whole. Is it not, in fact, legalized expropriation of a social asset by a small number of persons? If there can be a question of profit, should one not aim at collective social profit, in the common interests of the whole population and of the generations to come?

The experience of privatization in the United Kingdom strengthens the relevance and validity of these questions. Expropriation of a common social heritage by a small number of persons (the shareholders of the private companies) is particularly striking in Britain, where the profits have been so high that Tony Blair – who has no intention of questioning the privatization of water – imposed a special windfall tax in 1997 on 'excess profits'. The water companies were condemned to pay £1.6 billion in 1998 and 1999 – which shows how assiduous the company directors were in securing profits and handing out dividends, instead of making the promised investments to improve infrastructure and the quality of the water supply.[70]

Since privatization, the amount of waste due to leaking pipes has risen to 30 per cent. Interruptions of the supply are a common occurrence, even though prices increased by 55 per cent between 1990 and 1994. Disconnections for non-payment of bills have also become more frequent, with the number of people below the poverty line having soared from 7.4 million in 1979 to 13.9 million in 1994 during fifteen years of Conservative government under Margaret Thatcher.[71]

Those who defend the French-style water privatization argue that a distinction should be drawn between ownership and management: the French model is one of privatized management of water services, while the actual ownership of water remains in public hands. This point has only the appearance of correctness, however. For although the distinction is still formally in place, in practice ownership has been swept away in any real meaning of the word.

When ownership is really public, the owner (that is, the nation acting through its elected representatives) is master of the asset and of the use to which it is put, determining both its value and its utility functions. But the French experience shows that private water companies have gradually ejected politics from the decision-making process concerning norms, standards, prices, and so on. They have been able to do this thanks to the scientific and technological mastery and the managerial, financial, legal and other skills they acquired when they became 'the ones in charge'. Once water management leaves the public sphere, all the elements of control are transferred to the private sector.

Thus, like the landlords of the Middle Ages, the money lords in privatized water regimes try to draw the maximum short-term profit from the natural resource whose ownership has been allocated to them by human laws. Now, a large number of persons, as well as economic, scientific and political organizations, push for more and more human laws to transfer the ownership of water to money lords. These particular money lords will not behave any differently from others in the past or present: they will apply a logic of expropriation based not upon solidarity but upon the principle that 'the most competitive wins'. The time has come to stop allowing these human laws to pile up – the more so as they largely reflect a particular culture and approach, that of the technology lords.

Technology lords

To show what is meant by 'technology lords' (as one variant of the water lords defined at the end of Chapter 1), let us look at two cases: dams, and bottled water.[72]

The dam builders

Dams have existed ever since humankind discovered their usefulness as a means of protection, water diversion and riverflow control. Considerable progress has been made in the design of more solid, efficient and cost-effective constructions, sometimes amounting to great feats of skill and ingenuity on the part of *homo faber*. In recent years, *homo faber* has been hoist with the petard of his own achievements, as these have fuelled ever more grandiose ambitions and megaprojects. One has the sense that he does not wish to draw the much-needed lessons of past and recent experience, which would show the limits and perverse effects of a logic dominated by faith in the power of *techné* (technocracy).

In the world today there are roughly 40,000 large dams.[73] Of these, 35,000 have been built since 1950, and the rate of construction has increased in the past fifteen years. They have spread everywhere, with the largest number concentrated in China, followed by the United States, the former Soviet Union, Japan and India.[74] Large dams – like skyscrapers, aircraft carriers, 500,000-ton ships, cloned animals and nuclear power stations – have become one of the most symbolic modern forms of the mastery of nature by means of technology. They are regarded as the work of giants, animated by a superhuman will and by visions capable of 'crossing geography and history' by reshaping nature and human life.

Nature and human history have demonstrated, sometimes quite dramatically, the fragility and limits and also the irrational side of these works. First of all, the scale of the risks increases. The more enormous the dams, the more they are both cause and object of what since the 1970s have been called *major technological risks*.[75] A nuclear power station, an airport, an urban electricity network, a chemicals factory, a gigantic oil tanker, a train carrying toxic material: all these are now major technological risks, in the sense that, were they to break down in an accident, the devastating consequences for human beings (and human lives), for nature (through pollution/contamination over vast areas for a long period of time) and for society (economic costs, dislocation, moral damage, destruction of the social fabric) would reach proportions never experienced before. It is hardly necessary to recall the devastation that accompanies the bursting of a large dam. The catastrophic floods of August 1998 in China are in many respects not unconnected with the large number of dams for which China has shown a special fondness. The fact that public opinion in the United States, Switzerland, Germany – among other countries – has been very active in the protest against the construction of the massive Three Gorges dam suggests that people are more and more aware of the need and possibility for a new model of development based on sustainable technology and goods.

The second set of drawbacks concern the direct and indirect costs of the dams. As a general rule, their champions tend to underestimate the costs and risks and to overestimate the benefits and technological reliability, especially with regard to the immediate impact on people living in or close to the zone of the future basin. Whole villages are evacuated, sometimes thousands forced to leave their homes. It has been estimated that between 30 and 60 million (10 million in China) have been displaced throughout the world as a result of dam construction.[76] Financial or other compensation cannot make up for the social, psychological and economic losses. Resettlement in new villages or urban districts always involves major traumas, as the people in question cannot easily resume their former way of life and economic activities (whether in agriculture, fishing or crafts). The effects on the ecosystem may also be irreversibly disastrous: shrinkage of biodiversity (extinction of migratory fish and aquatic plants), reduction of the downstream water supply, blockage of sediment often indispensable for agricultural activity, and the rising of aquifer layers.

Third, large dams turn out to disappoint the high hopes invested in them. Quite often the newly created basins constitute a danger to health, as the villages, forests and flooded agricultural areas become a source of pollution and putrefaction; the stagnant, polluted waters bring diseases such as malaria and schistosomiasis; and the costs of maintenance and purification rise sharply. Moreover, the accumulated water and the hydroelectric generation prove to be less than anticipated, and local people find they have less access to drinking water and energy, which have been earmarked for urban populations perhaps hundreds of kilometres away or for industrial and intensive agricultural complexes. The traditional community systems of irrigation and water supply management are also destroyed beyond repair.

Finally, dams built in remote areas such as the Canadian Great North often have a negative impact on indigenous peoples and communities. In Lesotho (just one example among many), the construction of the Katse dam has brutally dislocated the lives of local people and led to the spread of new diseases such as AIDS.

In various countries around the world, an awareness of all these effects has given rise to critical movements opposed to large dams. Campaigns have been waged for the building of smaller dams, as part of a more integrated, sustainable and participatory policy for water, agriculture and energy. Among their successes has been the cancellation of a large project in India (the Narmada dam), which the World Bank had championed for more than seven years and helped to finance. The inspection panel set up by the Bank to make an independent assessment of the project eventually found in favour of the movements opposing it. It is to be hoped that clear lessons for the future will also be drawn from the devastation caused by another gigantic project – the Birecik dam on the Euphrates in Turkey – on which work began in 1996. Its construction will involve the disappearance of 31 villages (out of 44 seriously affected) and the partial submergence of the town of Halfeti, with considerable loss of its architectural and urban heritage. The economic and social life of three of its districts will be completely altered.

No one should underestimate the creative, innovative force underlying the technological and human challenge of large dam construction. It will probably never be possible to eliminate the human Promethean will to go ever further in defying the forces and limits of nature and of mankind.

More prosaically, it is an undeniable fact that since the Second World War the demand for large-scale public works, including dams, has become an important link in the chain, a highly profitable sector for civil engineering companies, electricity corporations, and firms specializing in finance, consultancy or logistics. In recent years this has been especially the case with construction in the underdeveloped countries.

The main beneficiaries of the large-scale works funded by the World Bank (or its regional subsidiaries) and the IMF – within the framework of UN development aid programmes for the Third World – have been the multinational corporations originating in North America, Western Europe and Japan which financially support the World Bank and the IMF. Not only do local populations often derive fewer benefits from such operations than the construction, management and consulting companies; they also

come out of them with a higher burden of debt than before. All
told, the companies of the developed countries gain greater
wealth than local populations from the construction of large
dams.

The water bottlers

The case of bottled water throws light on other aspects of how the
technology lords view the world. Here the virtues of technology
are applied to one of the deepest and most significant myths of
contemporary humanity: the myth of perfect health. Whether a
passing fashion or not, this has been made one of the priority goals
of Western societies, linked in a way to the overcoming of death,
and expressed in a systematic war upon everything that might
damage perfect health. Pure water has become a major issue in this
trend. The water bottlers' publicity not only highlights the
'natural' character of their water (self-evident because it comes
straight from 'natural', not yet polluted, sources), but even suggests
that it is more natural than nature itself. 'Man-made' water is
claimed to be purer than anything nature can offer – which is
indeed a theoretical and practical possibility – and as such con-
tributes to the goal of perfect health. Its advantage over tap water
is no longer linked either to its taste or to the medically therapeu-
tic properties of spring water, but rather to its biological quality
and diversity.

The bottlers proudly record the development of what is known
in the business world as 'customization'. The product they offer is
not just pure water but specially pure water adapted to the highly
diverse needs of (well-off) consumers: water for sportsmen and
sportswomen, water for pensioners, water for pregnant women,
water for babies, water for growing children – all part of the contribution of bottled water in general to perfect health at every age
and in all circumstances. The emergence of 'all water' restaurants
(which serve all kinds of water from distant lands, in bottles

made of various materials and in various shapes for collectors to add to their collection) would appear to be part of this same phenomenon.

The new synthetic water created by Nestlé laboratories probably also falls under the logic of a quest for 'perfect water' adaptable to the health requirements of different sections of the population, as well as being an attempt to boost Nestlé's reputation in health matters after the setbacks of its powdered milk in Africa. Nor is Nestlé the only company in a field that promises to be highly lucrative, especially in the underdeveloped countries where the commercialization of purified water – even through *ad hoc* street fountains – is becoming more and more widespread. Nestlé remains the world number one in bottled water, with brands such as Perrier, Contrex, Vittel, Volvet and San Pellegrino; but Danone (number two, with Evian, Volvic, Frarrarelle, Cannon, Villa del Sur and others in its portfolio), as well as Coca Cola and Pepsi Cola, are also growing presences. Coca Cola, for example, is actively marketing a purified sparkling water called Bonaqua in more than thirty countries. Water treatment companies such as Suez–Lyonnaise des eaux and Vivendi are also showing increasing interest.

In short, no one knows what the water industry has in store for the future. It is quite clear, however, that if the future is left up to them (and to their logic of profit and competition), we will see ferocious new battles among freshwater distribution companies, multi-utility corporations (combining water, gas, electricity, telecoms and public works), water bottlers and fizzy drinks producers. The global water landscape is at risk of major transformations, as a result of mergers, takeover bids and interlocking equity interests dictated mainly by the power logic of finance and the market, for which the right of every human being to have access to water is a matter of secondary and quite subordinate interest.

Hopes are sometimes pinned on desalination technology;

experts were even predicting as long ago as the early 1960s that water produced in this way would become competitive by the year 1980. It is an old dream of humanity: to transform sea water into drinking water. And indeed the promising signs are continuing to multiply. Some say that the cost of a cubic metre of desalinated water has fallen from $10 or more a decade ago to $1.50 today. Others, basing themselves on the results of the experimental plant at Huntington Beach (California), jointly sponsored by the Metropolitan Water District of Southern California and the Technion-Israel Institute of Technology, insist that the cost has already fallen as low as 60 cents a cubic metre. But constantly falling prices are not enough for the desalination of sea water to become widespread in the next fifteen to twenty years and thus to help solve the water supply problem in the many countries faced with shortages.

Apart from the strategic security argument (which, for example, is likely to convince Israel to abandon the desalination solution and to prefer a long-term contract for water imports from Turkey), two serious obstacles remain to the advance of desalination: one is that the production of hundreds of millions of cubic metres of desalinated water would require enormous quantities of energy, thereby increasing carbon dioxide emissions into the atmosphere and worsening the phenomenon of global warming; the other is that the dumping of residual waste water at high temperatures into the open sea would be a major source of marine pollution.[77]

Technology can help, but it cannot guarantee the right to water for the 1.4 billion people around the world who are today deprived of it. Nor is it the means to solve the interstate conflicts that threaten to become more widespread and acute. The path to real solutions passes through the political, cultural and socio-economic changes that we described in Chapter 1 as 'the first revolution of the twenty-first century'.

CHAPTER 3

BUILDING
A DIFFERENT
FUTURE

The World Water Contract

We have seen in the previous chapters why there is a *water problem* in the world, despite the scientific and technological progress of the past fifty years and the deployment of economic and financial efforts to solve it. It is a problem *for the world*, however much it may differ according to a country's standard of living (between Scandinavia and India, for example), to the character of a particular geographical zone (the Sahel and Quebec), to the composition of social groups (rich and poor), or to the nature of water usage (irrigation, industrial activity, household consumption).

We have also seen that, because this problem affects the bases and the sustainability of relations between human societies and the ecosystem Earth, it directly or indirectly concerns *everyone* in the world. Just as no one can say they are untouched by the nuclear problem, so – *mutatis mutandis* – no one can say individually or collectively that the water problem does not concern them, even if they are swimming in an abundance of the highest-quality water.

Now at last we can understand why the starting point of the World Water Contract is the *eight billion people* who will live on earth twenty years from now, including more than three billion – if present trends persist – who will not have access to safe drinking water and five billion who will face major problems of scarcity and quality. The World Water Contract is one of a series of actions undertaken by numerous groups, movements and international bodies to ensure that the trends do not persist.

The three critical situations we face

Today the water problem involves *three major critical situations*:

1 Non-access to a sufficient quantity of drinking water for 1.4 billion
   people and to water of sufficient quality for more than 2 billion

Of great importance for the World Water Contract is the acute accessibility crisis for people living in the 650 cities that will have a population of more than a million by the year 2020 (especially the 600 of these in the poor countries). Previous chapters have analysed the main causes of this non-access: rapid and chaotic growth of the cities, the prioritization of such objectives as military spending, the profit logic or the consumption needs of the ruling classes, instead of investment in basic infrastructure to improve the water supply situation.

2 The destruction/degradation of water as a fundamental resource of
   the ecosystem Earth and of human life

We have shown that, among the many factors in play, the following have played the decisive role in the processes of water destruction/degradation:

- irrigation practices in intensive industrialized agriculture (high level of waste, salination of underground water);

- pollution/contamination caused by industrial activities and the still largely inappropriate and inadequate forms of management (or non-management) of urban waste;

- excessive use of water resources due to inefficient systems of water production/distribution and consumption;

- thoughtless multiplication of 'large dams' (more than forty thousand around the world);

- long-term effects of major natural catastrophes resulting from human action (drought, floods, landslides, burst dams, etc.).

Two crucial aspects that have not received the same attention in recent years as industrial pollution and the management of urban waste should definitely be among the priorities to be addressed: namely, waste and pollution associated with intensive agriculture and irrigation, and the construction of large dams.

3 *An absence of worldwide rules, and of people to sustain a water politics based on solidarity, at a time of glaring structural weaknesses and defects in local water authorities*

Although water is increasingly recognized as a vital good, and although access to it for all has been formally declared an objective by the United Nations and other international governmental or private bodies active in relation to water, this has not given rise to a world corpus of basic laws beyond the framework of national states. World water legislation is dramatically lacking. The successful efforts since the International Decade of Drinking Water and Sanitation to get two major conventions drafted and signed should be positively welcomed, but there is a deplorable lack of world bodies with sufficient powers to provide a clear sense of direction and to monitor the implementation of existing conventions.

By contrast, there are signs everywhere in the developed countries that the state and local communities (towns, villages, basins, regions, indigenous communities) have been losing control over what happens to water (production, distribution, sanitation, conservation). As we have seen, water is more and more being transformed from a public asset (*res publica*) into an economic asset, its ownership and management guided by the logics and practices peculiar to the capitalist market economy.

Among the reasons for this, let us recall:

- the principle of state sovereignty in the ownership and use of water resources – a principle which, in its extreme form of

absolute territorial sovereignty, has been at the root of most
water wars between countries;

- the deterioration of public finances, especially at the level of
  local councils, where debt is becoming a major obstacle to
  their capacity to manage public assets;

- the growing abdication of responsibility by public authorities
  (governments, parliaments) in favour of private subjects (especially multinational corporations and international financial
  bodies), in respect of resource allocation and the distribution
  of wealth thereby created;

- the successful pressure for water privatization.

Three top priorities emerge from this for the World Water
Contract: (1) fundamental work on constitutional law (world water
legislation) to facilitate (2) effective action for 'peace through
water' and (3) the introduction and/or promotion of democratic
management of water by local communities, involving, among
other things, the creation of 'water parliaments'. The table on page
89 presents a summary of the analysis above.

Towards the World Water Contract

Faced with this set of critical situations, one can hardly be content
with pragmatic, 'realistic', partial and uncoordinated responses, or
with reductionist or simplistic global visions such as the ones we
examined in the last chapter ('Just give the economy its head!';
'Just set the fair market price for water!'; 'Just apply the "user
pays" principle!').

The most-needed responses will involve:

- respect for *new rules* reflecting a veritable revolution in ways
  of looking at water and at water-mediated relations among
  human beings;

World water problem: the three most critical situations

Situation	Main causes	Area of key priority
• Non-access to drinking water for 1.4 billion (quantity), more than 2 billion (quality)	• Rapid population increase • Chaotic growth of cities • Prioritization of other objectives (military, profit, ruling-class lifestyle)	• Access to water in 600 cities with population of 1 million plus by the year 2020, located in Africa, Asia, Latin America, former USSR
• Destruction/degradation of water as fundamental non-substitutable resource of ecosystem Earth and of human, animal and vegetable life	• Irrigation in intensive industrial agriculture • Pollution/contamination by industrial activities and urban waste • Excessive abstraction of water due to wastefulness • Thoughtless multiplication of large dams	• Changing irrigation water systems • Questioning usefulness of large dams
• Lack of worldwide rules and weakening of local authority direction and supervision of water as common social heritage	• National(ist) geoeconomic and political strategies (climaxing in absolute territorial sovereignty) • Growing abdication of responsibility by public authorities (governments, parliaments, local councils) for ownership and regulation in favour of (increasingly multi-national) private subjects • 'Crisis' of public finances, especially at local level	• Drafting and implementing world water legislation • Creating bodies to exercise democratic decision making and supervision at global and local levels (water parliaments) • Peace through water

Source: R. Petrella (1998)

- the development of *new means*, expressed in ways of managing water that are capable of rebuilding solidarity at the level of local communities, across different communities and generations, and that are sustainable in terms of the ecosystem Earth.

It is precisely the function of the World Water Contract to set in motion a process which, over the next fifteen to twenty years, should make it possible, on a basis of cooperation and solidarity, to eliminate the causes of the three major critical situations that make up the world water problem.

Far from being a deed signed, sealed and delivered once and for all, as some might imagine it to be, the World Water Contract is a dynamic not free from conflict, controversy and revision. It could never be the umpteenth 'enlightened' project of a global elite watching from its glittering tribune of twenty-first-century modernity over the future welfare of the world and offering a new prescription for the people to apply. As we shall see, the World Water Contract will be participatory and associative or it will not be at all. This is why such a high priority will be given to parliaments (sites where all members of a human community are represented) and the structures of direct democracy (local community ownership and management of the water heritage), as well as to the action of social movements, the partnership of industry, finance and civil society, and initiatives to deepen solidarity that should be taken by banks, insurance firms and even industrial companies.

The Water Contract is a tool of innovation and change. It is not designed to prioritize the reasoning and interests of this or that social group, this or that community, this or that continent. On the contrary, it is designed to give a global community the right to exist, to assert the inescapable need for a global politics of solidarity among all human beings on the planet.

This is precisely why, in the opening pages of this Manifesto, we called for the first revolution of the twenty-first century.

The founding principle: water as a common global good

The World Water Contract is based upon the recognition of water as a vital common global heritage.

As we saw in the last chapter, water is not like other natural resources. No alternative can substitute for it, and so it is more than a resource: it is a *vital asset* for every living thing and for the ecosystem Earth as a whole. Every human being has the right, individually and collectively, to have access to this vital asset. Access to water and the obligation to conserve it for the purposes of survival pertain to humanity collectively; they can never be the object of private individual appropriation. The use and conservation of water are a result of human history, with its legacy of knowledge, practices, tools and organizations upon which no individual can claim to exercise private ownership rights. Hence the character of water as a common patrimonial asset. The conditions and means of access to water and its conservation are also not an individual matter but a task and responsibility for all human beings together − which reinforces the character of water as a common patrimonial asset. But if water is a vital asset and a common heritage for every local human community (village, town, region, country), the forms, conditions and means of its use are bound up with its character as a vital asset for the ecosystem Earth as a whole. *The primary subject of the common water heritage is thus the worldwide human community − hence the character of water as a global asset.*

All these features make of water a *social asset* and by antonomasia a *planetary asset*. This explains why the intrinsic finality of water (the way it is conceived and looked upon) must be one based upon *solidarity* and *sustainability*.

In accordance with the founding principle, each organized human community has the right to use water for its vital needs and for the social and economic welfare of all its members. At the same time, it must guarantee access and use to other human communities which share or do not share the same aquiferous basin, according to the modes of solidarity and sustainability agreed for this purpose (contractual approach).

The inalienable rights and duties with regard to water are collective rights and duties, *not individual or private ones. They belong to the whole of the world's population. Control and supervision of priorities in the exercise and enjoyment of these rights and duties should take place at the level of each human community, on behalf of and as a trust from the rights and duties of the world human community, which remains the primary subject of the common water heritage.*

At the present time, the subjects recognized as juridical subjects of rights and duties are individuals, public or private organizations, national states and international or intergovernmental organizations. The world human community as such is not, or is not yet, a subject to which rights and duties are attributed. This explains the capital importance of the Contract and its founding principle, as well as the urgent need for a body of world water law that will give birth to global subjects responsible for water who might then be capable of ensuring that the rights and duties of each local human community are respected better than they are at present. The principal duties in relation to water, which are also in the interests of future generations, are an obligation to ensure that it is conserved in harmony with sources of natural renewal and that its quality is properly maintained.

Let us now look at the application of this principle in the specific case of Quebec. In June 1997 the government of Quebec declared water to be 'an asset of the Québécois people' and gave up any plans, at least temporarily, for its privatization. It was a decision of great political and symbolic importance. At the heart of the debate was the question of water sales outside Quebec – perhaps to New York and especially to California. The government asserted the property rights of the Québécois people to 'its' water, maintaining that if any were sold abroad the whole people should benefit from the proceeds in the shape of investments allocated to collective social and/or economic objectives. It is certainly a legitimate approach. It means that the water of Quebec is a vital common heritage of the whole people, which exercises its use rights and performs its duties of conservation/quality protection on behalf of and as a trust from the world's population and future generations. In the event of sales outside Quebec – the conditions of solidarity and sustainability having previously been respected – some of the proceeds would have to be placed at the disposal of a world water fund, to be used under the responsibility and supervision of a representative world authority (a water parliament).

Thus water, as a vital common heritage of humanity, cannot be the object of traditional commercial transactions across borders, or of acquisition by foreign investors. Water should be excluded from any convention or treaty signed under the auspices of the WTO, and from any treaty or agreement concerning the regulation of financial investment throughout the world. On the other hand, it is high time that water is regulated and protected by a legally binding world convention.

Principal objectives

The World Water Contract is inspired by *two principal objectives*.

The **first** objective is *basic access to water for every human being and every human community*.

Basic access for *every human being* means that he or she can enjoy the minimum quantity of fresh drinking water that society considers necessary and indispensable to a decent life, and that the quality of this water is in accordance with world health norms.

Basic access *for every human community* means that it can

consume the quantity of water necessary and indispensable for the needs of local economic and social development, either by using locally available resources or by sharing, on a basis of solidarity, water available in other regions near or far.

Basic access must be recognized as *an inalienable fundamental political, economic and social right, at once individual and collective.*

In order to achieve this, at least on the plane of legislative provisions, it will be necessary to sensitize and mobilize public opinion through *national and international campaigns* with the following aims:

1   The drafting and approval of *a legally binding world water convention* which integrates the political, economic and social right of individual and collective access to water into the Universal Declaration of Human Rights.

The same approach will have to be applied to other charters and conventions bearing upon human rights, national rights and minority rights. This will mean enriching and expanding international conventions already signed on water, such as the Strasbourg and Paris Declarations (see page 25).

The need for a world water convention to establish adequate forms of water management on a world scale has been vigorously endorsed in *Implementation of Agenda 21 in German Water Resources Policy*, a report presented by the German NGO Forum on Environment and Development at the Sixth Session of the UN Commission on Sustainable Development.[78]

2   The *modification of existing national constitutions or water laws*, or the passing of a new water law. In the case of Europe, the European Parliament should work with the European Commission to introduce an amendment to the framework directive on water, with the aim of building into it a recognition of water as a vital common world heritage and the right of individuals and collectives to have basic access to water.

Achievement of this first objective will involve a clear and precise definition of the system for determining the costs of basic access for all and how they will be met. We have already seen that the use of market mechanisms to set a price for water – even if, as in France, it is theoretically the public authorities which set it – does not constitute a solution. Moreover, the idea of a single system of cost calculation and price determination for every community and country in the world, with market price as a supposedly neutral device valid everywhere, should be most energetically resisted.

According to the Water Manifesto, the costs of providing access to water, and of purifying and conserving it, should be met collectively through a 'pricing' system that might be based on the following principles.

The local human communities responsible for ensuring that priority is given to the exercise of individual and collective rights and duties, on behalf of and as a trust from the world community, should determine for a period specified by themselves the total cost to the community of basic access to water. Using this as a guide, and taking into account the cost of other necessary prior and subsequent activities, they should then set the prices for freshwater distribution and sanitation services. It is proposed that they should adopt for this *a 'pricing' system graduated in accordance with principles of solidarity and sustainability*.

The first tier of this graduated system would apply to the minimum volume/quality annually necessary and indispensable for each human being and each human community. This basic amount would not be billed either individually or collectively, on the grounds that 'access to basic water is not a matter of choice'. It would be up to local communities to decide whether each household (adult person, couple, etc.) should by way of exception be asked to pay a general contribution (a kind of direct tax) for the financing of the collective costs. But public or private organizations, communes, villages and companies would be required to make a lump sum payment to the community.

The second tier would relate to any use of water over and above the allocated vital minimum. The unit price of such consumption would itself increase quite steeply after a certain threshold of sustainability had been exceeded. Beyond certain limits considered improper and intolerable by the community (the third tier), the user would be subject to punitive sanctions, as is the case today with a car owner driving at 120 or 130 miles an hour on the motorway (withdrawal of driving licence, heavy fines, negative endorsement of character reference, etc.). This would mark a change, a necessary change, from the 'user pays' or 'polluter pays' principle. Of course industries (and some commercial activities) which consume a huge amount of water should be taxed more than others and be prohibited from passing on such taxes in their prices. But the right to water for all, and the individual and collective obligation to conserve and share this vital common heritage, entail that no one should be allowed, just through paying more, to use water in an irresponsible and unjustified manner. That should be declared illegal. And without falling into the excesses of a public criminalization of water consumption (the populist, fundamentalist and dictatorial aberrations would be too numerous), it is crucially important to foster a culture of reasonable management of the water heritage.

Proposals of this kind may be found nearly everywhere. In Belgium, for example, the NGO Coordination gaz–éléctricité–eau calls for a 'tripartite pricing system: a fixed charge kept as low as possible, unit prices as low as possible for the first block of consumption corresponding to what is considered a fundamental right (perhaps with annual rebates per cubic metre for certain categories of household), and rising price levels for everything else'.

The costs of basic access for all and of purification/conservation/quality-maintenance would thus come out of the collective budget. This would certainly require a better, more efficient local and international struggle against fraud and tax evasion (growing trends in the age of globalization and deregulated capital movements), as well as more progressive and redistributive tax rules based upon principles of solidarity.

This is not the way the wind is blowing today – which is not good news for social justice, economic efficiency, political democracy and ecological survival.

The second principal objective, as defined above, concerns *integrated sustainable management of water in keeping with principles of solidarity*. This implies a *threefold duty*, both individual and collective, in the use and conservation/protection of water.

The first duty is one of *solidarity* with other human communities which, for one reason or another, are temporarily or structurally in a situation of water shortage or scarcity. It should be possible for mutual contracts to be signed and implemented among various communities, in accordance with the agreed principles and content of the *world water law*. Such mutual contracts would, for example, regulate inter-community relations at the level of a river, a lake or a basin, or even among countries quite distant from one another, within the framework of wider cooperation and solidarity in every area vital to the communities in question (agriculture, health, energy, etc.). The practice of river contracts, which is quite widespread in Belgium, France and Switzerland – like the forest contracts encouraged at an international level – goes only part of the way in the direction outlined above, since it does not cover the distribution and sharing of water. The time has come to take another step forward.

The second duty is one of *consistency* with the rights and freedoms one has granted to oneself: that is, not to act in such a way as to reduce or endanger the rights and freedoms of future generations, but to pass on to them the vital common heritage in conditions, if possible, better than those handed down by previous generations. The application of this duty should, among other things, prompt us to define an *index of security/health/sustainability* – as we have done to measure air pollution, global warming, beach pollution or salination of underground water, and as the United Nations Development Programme has done in developing those precious and extraordinary instruments of measurement: the Human Development Indicator (HDI) and the Poverty Indicator (PI). The task – first of all for the scientific community – should be to synthesize a global/local indicator of security/health/sustainability with regard to water, which would serve as a source of information and rigorous quantitative and qualitative analysis so that more coherent choices can be made for the future. Perhaps this is something else that could be carried out by the United Nations Global Environment Monitoring System.

The third and final duty is one of *protection/respect* of the ecosystem Earth. Chapter 18 of Agenda 21[79] has already said and proposed nearly everything on this matter (see in particular Section C, 'protection of water resources, water quality and aquatic ecosystem'), and it would be enough to implement its detailed action programmes for this third duty to be satisfactorily fulfilled. We should remember that it was approved and signed by more than 130 heads of state (or their representatives).

The priority targets

In the light of what has been said above, the implementation of the World Water Contract should begin around four *priority targets*:

1 *Three billion taps by the year 2020*
The aim here is not only to prevent the number of people without access to drinking water from rising from today's 1.4 billion to 3.2 billion in the year 2020, but actually to bring the figure crashing down to 'zero water absence'. The word 'tap' might conjure up a

system centred upon access to water at the level of family house-holds (by 2020 the urban population will represent more than 65 per cent of the world total). But it also means taking account of present and future situations unlike those of today's Western urban settings (nuclear families, private individual housing units). 'Three billion taps' is not meant to imply an individual approach, even if the idea of a tap might suggest this. Basic access to water – as we have stressed a number of times – is a collective process. We live *together* with others!

2 Peace through water

More initiatives need to be taken to *defuse water conflicts*. Public opinion is not sufficiently informed or focused on the issue. The present 'water wars' are, as we have seen, part and parcel of the wider conflicts between states; 'peace through water' will depend upon a wider peace. Nevertheless, two major social and political forces could and should play a decisive role in the defusing of water conflicts:

- *Community movements and NGOs linked to the world's main religious currents* (Buddhist, Catholic, Muslim, Protestant, Orthodox, Shinto, Jewish, Hindu, animist, etc.) are still too reticent. They should together exert much stronger pressure on their churches and institutions to place their weight and authority behind the goal of 'peace through water'.[80]

- *Parliamentarians* could do more to promote the general interest of the peoples of the world, and do it better than the state representatives who have signed universal, continental or national declarations and charters on human rights, peace, minority rights or sustainable development.

3 Reducing waste, changing irrigation, saying no to large dams

There is a broad consensus among the parties concerned that it is urgently necessary to halt the damage caused by irrigation systems tied to intensive industrial agriculture. The problem lies in the weakness of national public authorities, which seem incapable of coordinating their efforts internationally to impose on agribusiness the rules, norms and standards of a socially and environmentally sustainable agriculture, and thereby drastically to reduce the extent of water leakage, pollution and waste. Solutions such as drip-administered irrigation should be generally promoted to face the urgency of the tasks, but the key reforms will have to concern the system of agriculture itself.

Equal urgency is required in addressing the problem of large dams. An ever greater number of experts, social movements and local communities are in favour of *a worldwide moratorium* on the construction of large dams. Here, too, there is an alternative worth exploring: namely, the construction of smaller dams mainly geared to the production, conservation and distribution of water for local agricultural, industrial and urban uses.

4   *The 600 cities of Latin America, Asia, Africa and the former USSR whose populations will be over a million by 2020*

More than two-thirds of the 8 billion people living on earth by the year 2020 – including more than a billion for whom bare survival will be life's main problem – will reside in the 600 or more 'millionaire' cities on three continents. These cities will not by themselves have the financial, technological or economic capacity to prevent them becoming ravaged by thirst, and prey to illness and poverty, to human, social and environmental degradation. So much has been said, and said repeatedly, by the UNDP, WHO, FAO, UNICEF, Habitat II and the World Bank.

What is the point of increasing the world's riches if a growing number of people know only ever greater poverty and find the right to life ever more inaccessible? What is the point of increasing the value of shares in Vivendi, Lyonnaise des eaux, Nestlé,

Danone, Severn–Trent, Thames Water, IBM, Microsoft, GM, Dresdner Bank, Citicorp, and so on, if the wealth produced throughout the world is mainly appropriated by those who already possess the largest share?

Numerous companies – for example, those involved in the Prince Charles International Business Forum for Responsible Care – stress their willingness to act in a 'responsible' and 'ethical' manner. Do they mean responsibility and ethical behaviour toward their shareholders, only now extended to a whole array of 'stakeholders' (a nebulous concept embracing company employees, clients and suppliers, consumers and public authorities)? Or do they mean responsibility and ethics redefined to take account of the real problems at stake in the next twenty years? We are prepared to bet on the latter. But then is not basic access to water and sanitation in those 600 of the world's poorest cities a clear-cut objective by which to measure the newly ethical and responsible approach of industrial and financial corporations in the developed world?

Making a reality of all this: what needs to be done

It is not enough to define, as we have done, the founding principle, the principal objectives and the priority targets. For the priorities to become credible, it is also necessary to specify the initiatives that need to be taken, the means to be used, and the social forces upon which the various proposals for action can depend.

The spirit informing the approach of the World Water Contract is *long-term* (solidarity across generations, sustainability of priorities for at least twenty years); it is *participatory* (we strongly believe in a cooperative quest for the best solutions in the general interest, not in competition that satisfies the winners' interests); it is *communitarian* (water rights and duties are individual and collective, local communities being the priority subjects for the management/conservation of the common heritage on behalf of and as a trust from the world community); it is *global* (the future of our societies is a common future; there is globalization not only of markets and finances; solidarity and sustainability go beyond rivers and the frontiers of aquiferous basins); and it is *progressive* (the aim of the priorities is to expand and strengthen the field of solidarity/sustainability among human communities and generations, at the global level of the ecosystem Earth).

The proposals for action assume the support, commitment and cooperation of four social actors:

- parliamentarians;

- community movements and organizations of civil society;

- scientists, intellectuals and the media;

- the trade unions.

These four actors should:

- put pressure on governments and key economic structures (above all, the big corporations);

- alert and mobilize public opinion (especially in the educational sector, at the workplace, among consumers, and in the web of small businesses);

- promote innovation at the level of law-making, values and economic and social practices.

Two actions have top priority from the point of view of time. The first is the establishment of a worldwide 'Water for Humanity' collective. Such a network, consisting mainly of community movements and organizations of civil society (including associations present in the business world), would have the task of promoting the definition, organization and implementation of *two worldwide 'water for humanity' campaigns*. The priority target of these campaigns will be 'three billion taps', which is also the first objective of the World Water Contract. The method of the campaigns will be the one shown to be effective by Amnesty International, Handicap International (against anti-personnel mines), Greenpeace, Oxfam and the 'Made in Dignity' campaign for fair trading – just to mention some of the best-known examples.

Already the issue of 'water for all' is on the agenda of such NGOs as Swissaid and Helvetas, the Foundation for Human Progress, the Worldwide Fund for Nature and many other local, national and continental organizations. To synergize, and thereby attain the critical mass necessary to be effective: this is the task facing the worldwide Water for Humanity collective as it seeks to build two campaigns whose specific forms may vary from one country to the next.

Within this general framework, it will be up to the Water for Humanity collective (that is, organizations belonging to the network) to launch the action around 'Living in the 600 cities of Africa, Latin America and Asia whose populations will be more than a million by the year 2020', which corresponds to the fourth priority target of the contract.

Of the partners with which a kind of cooperative contract should be reached in pursuit of this priority, the main emphasis should be on the world of banking and financial companies – above all, on the sector of credit and savings banks and people's cooperatives, whose culture has deep roots in the history of mutualist and cooperative movements. It is proposed that credit and savings banks and cooperatives should take the initiative in persuading all banking and financial institutions *to allocate 0.01 per cent of their daily international transactions to a 'Water Fund for 600 Cities'*. This fund, to be managed by the donors themselves, would have the sole purpose of providing joint public/private finance for infrastructural investment programmes and the production/distribution of drinking water and sanitation in the 600 cities in question.

This proposal, which involves a kind of self-taxation, does not differ in spirit from the World Water Council's idea of setting up a World Water Fund. But without being mutually exclusive, they do differ in their principles of organization, their forms and conditions of implementation, their principal actors and their short-term and long-term results.

The proposal is also different from the celebrated Tobin Tax (0.25 per cent levied on all international financial transactions), which is aimed at combating global speculation and tax evasion. Our own principle is that the donors decide for themselves. It takes seriously the directors of large industrial and financial corporations when they declare their openness to a new ethic and a responsible corporate development policy *vis-à-vis* the totality of 'stakeholders'. It goes beyond the forms of public/private water partnership currently proposed by the water companies, international financial bodies and various governments (the limits and ambiguities of which we briefly addressed in the last chapter). Our own proposal implies not only that the banks (in the broadest sense of the term) should carry out a partial and limited distribution of their valued-added to the benefit of the 600 world cities rather than their shareholders, but also that companies specializing in construction and distribution activities with regard to drinking water and sanitation should participate on a basis of cooperation and solidarity.

The second priority action should concern the establishment of *a network of 'parliamentarians for water'*. Initially comprising a representative core of parliamentarians from various countries around the world, such a network – in close liaison with the existing Globe parliamentary network – should take the following immediate initiatives :

- Publish an *open letter to the world* in the most influential of the international dailies, explaining why parliaments should commit themselves to 'water for all – a common world asset'.

- Organize a *conference of 'Peace through Water' parliaments*, as a contribution to the second of the Contract's priority targets. Such a conference might be held, symbolically, at the end of 2001, the first year of the third millennium, and then be integrated with the ongoing work on a millennium declaration that is being promoted and coordinated by UNESCO. Also symbolically, the conference might take place in Valencia, the location of the Water Tribunal which has exercised its jurisdiction there ever since 1492! One of the operational aims of the Conference of Peace through Water Parliaments would be precisely to promote the creation of a World Water Tribunal and a World Centre for the Monitoring of Economic and Social Water Rights.

It is hardly necessary to stress again the importance and the urgency of *arbitration in inter-state conflicts relating to the sharing and management of water resources*. This would be the chief function of the World Water Tribunal, for which Valencia would present itself as a prime location.

The creation of a World Centre for Economic and Social Water Rights corresponds to the need (underlined at the beginning of this chapter) for a body of world water law that would apply a legally binding world water convention. Such a world centre should especially benefit from the energetic support of public and private sector unions in the various continents of the world. In general, *trade unions* will have to play a major role in pressing for the priority targets of the World Water Contract, especially with regard to the promotion of water as a common social heritage and to the body of world water laws and the 600 cities. The task of the centre would be to carry out worldwide research and analysis for the development of a world body of water laws.

The campaign for a World Water Parliament should be launched as soon as possible; its initial phase should involve a kind of global public hearing on water-related issues. Held every three to four years on the initiative of one or several national parliaments (according to the principle of rotation and candidacy applied in such areas as the Olympic Games, the World Cup or the annual choice of a cultural capital of Europe), the hearing might last a week and bring together ten to fifteen thousand people. Its task would be to assess the state of water rights and duties throughout the world, to develop better knowledge and analysis of the problems, and to debate alternative scenarios for the promotion and strengthening of lasting solutions based upon principles of solidarity.

Within this framework, and more particularly in relation to the Peace through Water objective, the Parliamentarians for Water network might also encourage the creation of *parliamentary assemblies at the level of interstate basins* in cases where all other means of conflict resolution have proved ineffective. Here too, trade unions should be the spur to solutions based on solidarity.

Pragmatists will no doubt say that this is all just another fine programme. We are also convinced that it is a fine programme, and that it is realistic as well as idealistic. To get it started, we propose that the committee working towards a World Contract should change itself into a World Water Contract Association, with the task of taking to the baptism font, as quickly as possible, the Water for Humanity collective and the Parliamentarians for Water network.

CONCLUSIONS

The hour of democracy and solidarity

The control of water must be given to its true owners, to the inhabitants of planet Earth. It does not belong to nation states, nor to markets, corporations or shareholders. It belongs to human communities, from the smallest (villages) to the largest (the global community).

The Water Manifesto has shown, however, that it is not enough to proclaim that water belongs to humanity. This alone will not lead human beings to make reasonable, sustainable, efficient and cooperative use of it. In fact, the water problem is above all a problem of democracy and solidarity. If human societies, from grassroots communities up, are not driven by a democratic culture and practices based upon solidarity, then water itself becomes a source of social inequality and injustice.

Today water is a risk-filled domain. The risks are not only of shortage, pollution and waste, but once more (after a period when water was recognized as a public good and the risks were at least attenuated) of resurgent inequality, injustice, armed conflict and discord between human communities and generations.

It is high time that we humans learned to control the pumping, storage, production, use, conservation and protection of water, on a basis of democracy and solidarity and at every level of the organization of society. 'Good governance' in relation to water cannot be achieved except through democracy. To create the conditions for everyone to exercise their fundamental right of access to drinking water – which is a right to life – is a matter of citizenship.

The control of water cannot be left to the logic of finance and the market, for these guarantee the right to life only for solvent consumers and savers/property-owners/shareholders. For humanity to regain a right to life through and with safe water, both today and in the future, it is therefore necessary to invert the present trends toward commodification of every human activity and privatization of every good and service. The early part of the twenty-first century will be crucially important in this respect, and it will not be easy to bring about a change of course. Yet we can point to the abandonment of OECD-sponsored negotiations on the Multilateral Agreement on Investments following energetic public protests, as well as to the support of more and more political leaders for a revival of *politics* in monetary, financial, economic and social affairs at both a national and an international or global level. These are already signs that other paths and other regulatory systems are within the bounds of possibility.

The Water Manifesto: A Summary

The World Contract

Founding principle:
Water is a vital common
global heritage

Principal objectives

(1) Basic access to water for every human being and every
human community (an inalienable political, economic and
social right, at once individual and collective).

(2) Integrated sustainable management of water in keeping
with principles of solidarity (duty of individual and collective
responsibility to other human communities and the world's
population, to future generations, and to the
ecosystem Earth; principle of sharing, and
conservation/protection of water).

Priority targets for the next twenty years

In pursuit of Objective 1
• Access to water for
the world's poor population
(Three Billion Taps)

• Defusing water conflicts

(PEACE THROUGH WATER)

In pursuit of Objective 2
• Reduction of waste
(different irrigation,
a moratorium on large dams)
• Sanitation systems for the
650 cities whose populations
will exceed one million by the
year 2020/2025
(CITIES TO LIVE IN)

Summary of proposals for action

The first immediate action is to bring two networks into being:

1

GLOBAL 'WATER
FOR HUMANITY'
COLLECTIVE

(1) The global 'Water for Humanity' collective will organize two world-wide campaigns: (a) *Three Billion Taps*. The campaign should take place simultaneously in a number of cities. (b) *Living in Cities* – that is, the cities of Africa, Asia and Latin America projected to have a population of a million or more by the year 2020. To achieve these two objectives, it is proposed to engage the participation and commitment of the world of banking and finance (banks, savings banks, cooperatives), by establishing a fund for the 600 cities out of a 0.01 per cent tax raised by the banks and insurance companies on their own daily international financial transactions.

(2) The 'Parliamentarians for Water' network should:

2

'PARLIAMENTARIANS
FOR
WATER' NETWORK

- Publish a 'letter to the world' (a two-page summary of the World Water Contract) signed by 10 or 12 parliamentarians.

- Organize a conference of the 'Peace through Water' parliaments (in Valencia).

- Launch a campaign to establish a World Water Parliament, its members to be initially appointed by national parliaments.

- Promote the creation of parliamentary assemblies at the level of inter-state aquiferous basins, of which there are approximately 215 in the world.

- Support the creation of a World Water Tribunal and, with active trade union participation, a world forum/institute for economic and social rights relating to water. The forum/ institute should help in drafting a World Water Report and work closely with the World Water Parliament as well as those in charge of the world campaign for *Three Billion Taps*.

To set all these actions in motion, it is proposed to found a World Water Contract Association – actually a network, whose founding locations might include: Lisbon, Rio de Janeiro, Rabat, Los Angeles, New Delhi, Valencia, Montreal, Berlin, Buenos Aires, Paris, Brussels, Dakar and Tokyo. It would have both a framework function and a scientific-political function. The main activity of the Association, once the Collective and the Network have been launched, will be to publish every two years in close cooperation with the Centre for Economic and Social Rights, the World Water Report, to be funded from targeted public subscription.

NETWORK BASED MAINLY ON THE ORGANIZATIONS OF CIVIL SOCIETY

NOTES

1 See Group of Lisbon, *Limits to Competition*, Cambridge, Mass.: MIT Press, 1995; and Riccardo Petrella, *Le Bien Commun. Éloge de la solidarité*, Brussels: Éditions Labor, 1996.

2 This is, of course, to cut a long story very short.

3 See Denis Duclos, 'Naissance de l'hyperbourgeoisie', *Le Monde Diplomatique*, August 1998, pp. 16–17.

4 It is true that the top actors in pharmaceuticals and chemicals (if the new biotechnologies develop in a completely deregulated framework where it is possible to patent living organisms), and to a lesser extent in the agrifood complex, may also hope, and lay claim over the next 20–25 years, to be 'lords of the earth'. Everything will depend upon the overall conceptions (visions) and short-to-medium-term political choices prevalent in developed societies.

5 That the present-day 'information society' serves to heighten inequalities is now generally recognized as an empirical fact. There are those who, out of pragmatism or fatalism, accept this characteristic as difficult to remedy; and there are a significant minority who act to change it and to correct the prevailing trends.

6 On these questions, see F. Chesnais, *La mondialisation du capital*, 2nd edition, Paris: Syros, 1998; and Groupe de Lisbonne, *Le désarmement financier*, Brussels: Éditions Labor, 1999.

7 The work of the Club of Rome may still be read with interest, especially its first report that had such an impact: *The Limits to Growth*, New York:

New American Library, 1972. See also the reflections of Albert Jacquard, especially in *Voici le temps du monde fini*, Paris: Le Seuil, 1991.

8 On all these questions, see the report of the Brundtland Commission: *Our Common Future*, London: Fontana Books, 1988.

9 *International Herald Tribune*, 8 January 1998.

10 See United Nations Development Programme, *Human Development Report: Eradicating Poverty*, Washington 1997.

11 *International Herald Tribune*, June 1998.

12 See *Human Development Report, op. cit.*

13 Data presented to the UNESCO conference on 'World Water Resources' at the Dawn of the Twenty-first Century', Paris, 3–6 June 1998.

14 *Le Monde*, 17 January 1998, p. 11.

15 See 'Waterwars Ebb Away in West', *Financial Times*, 8 January 1998. On the Californian case, see the interesting article by Ujjayant Chakravorty and David Zilberman, 'Lessons of a Drought', *Down to Earth* (New Delhi), 30 June 1997, pp. 33–7.

16 An average person in the American South-West consumes a total of roughly 3,100 litres of water a day (for all purposes including irrigation), whereas the average domestic consumption of all Americans is 700 litres a day. In Belgium the average is in the region of 260 litres and in Italy 350 litres. In middle-income countries, on the other hand, it is approximately 150 litres, and in the towns of the Sahel it falls as low as 30 litres.

17 Special mention should be made of the World Bank's own work, edited by its vice-president I. Serageldin, *Towards Sustainable Management of Water Resources*, Washington, 1995.

18 See Stephan Schmidheiny (coordinator), *Changing Course*, Cambridge, Mass.: MIT Press, 1991.

19 The World Bank report on the state is a particularly instructive example of this. See *World Development Report, 1997. The State in a Changing World*, Washington: World Bank, 1997.

20 See the interesting work by Jean-Pierre Goubert: *La conquête de l'eau. L'avènement de la santé à l'âge industriel*, Paris: Laffont, 1986.

21 A thoroughly documented critique on this issue may be found in the work of the Delhi Centre for Science and Environment. See especially the report *Dying Wisdom: Rise, Fall and Potential of India's Traditional Water-Harvesting Systems*, ed. by Anil Agarwal and Sunita Narain, New Delhi, 1997.

22 We shall return in detail to these questions in Chapter Two. For an analysis of this statization of water for geopolitical and military reasons, see Jacques

Sironneau, *L'eau, nouvel enjeu stratégique mondial*, Paris: Éditions Economica, 1995.

23 See *Dying Wisdom, op. cit.*

24 A major instance of this has been the EU's Common Agricultural Policy; and the enthusiastic adoption of the 'green revolution' in Asia and Africa is due in part to the same reasons.

25 Already in the early 1970s, opponents of intensive agriculture assembled a remarkable and incontestable body of empirical evidence. Confirmation may be found in Fast, *L'utilisation des ressources agricoles et forestières en Europe: mise en question des modèles de développement prédominants*, Brussels: Commission des Communautés Européennes, October 1988.

26 See the reports of the FAO World Food Summit held in Rome in November 1996, and also UNDP, *Human Development Report*, 1997, pp. 39, 65–72, 125, 198–9.

27 Land degradation is one of the most serious structural phenomena involved in the mutation of ecosystems. The figures for the proportion of degraded arable dry land are very high: 73 per cent in Africa, 70 per cent in Asia, 54 per cent in Australia, 65 per cent in Europe, 74 per cent in North America and 72 per cent in South America. See Monique Maiguet, *L'homme et la sécheresse*, Paris: Masson, 1995; and United Nations Environment Programme (UNEP), *World Atlas of Desertification*, 1992.

28 See, in particular, the work sponsored and coordinated by the Foundation for Human Progress: 'Agriculture paysanne: une alternative à l'agriculture industrielle', *Passerelle* No. 4, 1993; and the proposals for a just and lasting land reform on which the Foodfirst Information and Action Network has been working in the last few years, in line with the principle that access to resources is a basic socio-economic human right.

29 United Nations, 'World Urbanization Prospects: the 1994 Revision', Database Population Division, New York, 1996. For an interesting study of several of the world's big cities, see François Valiron, 'Le cas des mégapoles', report to the symposium *L'eau et la vie des hommes au XXIème siècle*, UNESCO, Paris, 26–27 March 1996, organized by MURS and the Water Academy.

30 World Health Organization, *World Health Report 1996*, Geneva. Here is just one example: diarrhoea caused by polluted water affects some 500 million people every year and is the main cause of death among children below two years of age.

31 According to Jamal Anwar, some 38 million people in West Bengal and 50 million in Bangladesh (30 per cent of the country's population) are

exposed to the effects of arsenic-contaminated drinking water – a catastrophe which, in the author's view, is even worse than those of Chernobyl or Bhopal. See J. Anwar, *Arsenic Poisoning in Bangladesh* (n.d., n.p), a document received from the author on the occasion of a meeting in Valencia in late May 1998.

32 'Bataille planétaire pour l'or bleu', *Le Monde Diplomatique*, November 1997.

33 In order to avoid overloading the text, we have done no more than touch upon the problems of access to water in the rich developed countries, especially in urban areas. Although these have not reached the critical and dangerous level of Third World cities, the supply and quality problems (especially in relation to the treatment of waste water) are becoming more and more serious, as a result of the ageing of sewage systems and other infrastructure, or the inadequacy or even complete absence of purification plant.

34 See World Bank, *Water Resources Management*, Washington, 1993; P. Rogers, *Comprehensive Water Resources Management: a Concept Paper*, document 897, Washington: World Bank, 1992; and Elaine Geyer-Allély, *Water Consumption and Sustainable Water Resources*, Paris: OECD, 1998.

35 See World Water Council, *Newsletter* No. 2, December 1997.

36 On these points, the reader may consult Guy Le Moigne and Pierre-Frédéric Ténière, 'Les grands enjeux liés à la maîtrise de l'eau', *De l'eau pour demain*, special issue of *Revue française de géoéconomie*, No. 4, Winter 1997–8, pp. 37–46.

37 I. Serageldin, 'The Water Bomb', interview in *The Guardian*, 9 April 1995.

38 See Malin Falkenmark, 'Fresh Water: Time for a Modified Approach', *Ambio*, Stockholm: Vol. 15, No. 4, 1986, p. 92; and 'Regional Water Scarcity: a Widely Neglected Challenge', *People and the Planet*, No. 2, 1993, pp. 10–11.

39 Figures presented to the World Water Forum in Marrakesh, March 1997, and subsequently to the Ninth Congress of the International Water Resources Association (IWRA) in September 1997, on the basis of work conducted by the Stockholm Environment Institute and published in Commission on Sustainable Development, *General Assessment of World Water Resources*, April 1998. Of interest is also Igor A. Shiklomanov, *World Water Resources. A New Approach and Assessment for the 21st Century*, Paris: UNESCO, 1998.

40 *Ibid.*

41 Hervé Maneglier, *Histoire de l'eau*, Paris, 1992; Jacques Perennes, *L'eau et*

les hommes: essai géographique sur l'utilisation des eaux continentales, Paris: Bordes, 1977; *La guerre et l'eau*, symposium of the International Committee of the Red Cross, Montreux, 21–23 November 1994; Larbi Bouguerra, coordinator, 'L'eau pour tous', *Passerelles* No. 8, Paris, December 1995; *Les citadins et l'eau. Contrastes et similitudes à travers le monde*, survey conducted by l'Académie de l'eau on six of the world's largest metropolises, financed and published by l'Agence de l'eau Seine-Normandie, 1997; Paolo Sorcinelli, *Storia sociale dell'acqua*, Milan: Mondadori, 1998.

42 CONAIE, *Propuesta de ley de aguas*, Quito, 1996.

43 This discussion of Ecuador has largely drawn on the Swissaid dossier published on 6 February 1997 (Swissaid, rue de Bourg 49, 1002 Lausanne, Switzerland).

44 Here we shall limit ourselves to a few titles: Roger Cans, *La Bataille de l'eau*, Paris: Éditions Le Monde, 1997; Symposium *La Guerre et l'eau*, Comité International de la Croix-Rouge, Montreux, 21–23 November 1994; Jacques Sironneau, *L'eau: nouvel enjeu stratégique mondial*, Paris: Éditions Economica, 1995; United Nations, *Transborder Waters*, New York, 1997; Paul Samson and Bertrand Charrier, *International Freshwater Conflict: Issues and Prevention Strategies*, Conches (Switzerland): Green Cross International, 1997; Thomas F. Homer-Dixon, 'Environmental Scarcity and Violent Conflict: Evidence from Cases', *International Security*, Vol. 1, 1994, pp. 5–40; Peter H. Fleick, ed., *Water in Crisis: a Guide to the World's Freshwater Resources*, New York: Oxford University Press, 1993; Sandra Postel, 'Water Scarcity Spreading', in L. R. Brown *et al.*, eds, *Vital Signs 1993*, New York: W.W. Norton, 1993; Joyce R. Starr, 'Water Wars', *Foreign Policy*, No. 82, 1991, pp. 17–36; John K. Cooley, 'The War over Water', *Foreign Policy*, No. 54, 1984, pp. 3–26; Kent Hughes Butts, 'The Strategic Importance of Water', *Parameters*, Spring 1997, pp. 65–83; WMO/UNESCO, 'The World's Water: Is There Enough?', 1996; A Biswas, ed., *International Water Conflicts*, New York: Oxford University Press, 1994.

45 This is especially true of accounts that have appeared in the international press: for example, G. Pascal Zachary, 'As Demand Grows, Water Becomes a Global Concern', *Asian Wall Street Journal*, 6 December 1997; or Roger Cans, 'Devenue rare, l'eau risque d'être l'enjeu de conflits entre nations', *Le Monde*, 16 August 1995.

46 See the interesting work on the subject by Paul Samson and Bertrand Charrier, *International Freshwater Conflict, op. cit.*

47 David M. Wishart, 'The Breakdown of the Johnston Negotiations over the Jordan Waters', *Middle Eastern Studies,* No. 26, pp. 536–48. For more general analyses, see Environment and Conflicts Project (ENCOP), *Water Disputes in the Jordan Basin Region and Their Role in the Resolution of the Arab-Israeli Conflict,* Occasional Papers No. 13, 1995, Swiss Peace Foundation and Swiss Federal Institute of Technology; Aaron Wolf, 'Middle East Water Conflicts and Directions for Conflict Resolutions', report to the international conference *A 20/20 Vision for Food, Agriculture and the Environment,* organized by the International Food Policy Research Institute, 13–14 June 1995; J. A. Allan, ed., *Water, Peace and the Middle East,* New York: W.W. Norton, 1996; and Miriam R. Lowi, *Water and Power: the Politics of a Scarce Resource in the Jordan River Basin,* Cambridge: Cambridge University Press, 1995.

48 Sironneau, *op. cit.,* p. 40 (in the Italian version published by Arterios Editore, Trieste, 1997).

49 *Le Monde,* 28 March 1997.

50 See Uday Shankar, 'Disappearing Act', *Down to Earth,* New Delhi, 15 July 1993, pp. 25–30.

51 See Aniu Sharma, 'When A River Weeps', *Down to Earth ,* 15 April 1996, pp. 27–31.

52 Uday Shankar, 'Choking Slowly to Death', *Down to Earth,* 31 January 1993, pp. 25–36.

53 Access to these sources of information and analysis was made possible by the kindness of Sunita Narain, assistant director of the Centre for Science and Environment, which publishes the journal *Down to Earth.*

54 François Ramade, *Dictionnaire encyclopédique des sciences de l'eau,* Paris: Ediscience internationale, 1998, pp. 486–8.

55 The new water culture in the United States has introduced (or, for the time being, imported from Chile) the legal and practical possibility of 'water rights'. For example, a farmer who developed a more efficient irrigation system would be allowed to sell to an urban community water that he would not have consumed himself. This is hardly a surprising idea in a country like the United States, where everything is for sale – even 'the right to pollute'. In Chile this encouraged a particular kind of speculation by mining companies. Having received free from the state nearly all the rights over water at the moment of privatization, they today control the country's water market and have organized shortages to push up prices. See Larry D. Simpson, 'Les marchés des droits de l'eau aux États-Unis', in *De l'eau pour demain, op. cit.,* pp. 149–59.

56 An interesting and richly drawn critical survey of privatization, especially in Western Europe, may be found in Gérard de Selys, *Privé du public*, Brussels: Éditions Epo, 1996.

57 These changes have been analysed in detail in Group of Lisbon, *Limits to Competition*, op. cit., and Riccardo Petrella, *Le bien commun. Éloge de la solidarité*, op. cit.

58 On this precise point, see Riccardo Petrella, 'La ridistribuzione della ricchezza nel mondo. Il lungo cammíno del xxᵉ sècolo contro l'inuguaglianza', in *Enciclopedía del '900*, Turin: UTET, 1999, and as an essay published by Città di Castello: *L'Altra Pàgina*, 1999.

59 *International Herald Tribune*, 10 December 1997.

60 This is one possible reading of the state's devolution of its power of decision making and orientation in the global economy to the multilateral agencies. Another interpretation would maintain that the state only appears to be withdrawing, and that in fact the most powerful states profit from a strengthening of the multilateral agencies within which relations of force and inherently provisional compromises are the rules governing their action.

61 Gabriel Roth, *The Private Provision of Public Services in Developing Countries*, New York: Oxford University Press (for the World Bank), 1987.

62 The author mentions: a natural monopoly for the collection, purification and transport of water; decreasing costs; major externalities; billing difficulties; and the impossibility of disconnecting those who do not pay.

63 Banque mondiale, *Le secteur de l'eau au Maroc*, Report No. 12649-MOR, Washington, 1995, p. 25, quoted from Claudio Jampaglia, 'L'acqua e la città. Politiche e logiche di sviluppo in Marocco a partire dai progetti della Banca mondiale', thesis, Università degli studi di Milano, 1996–7, p. 272.

64 *Financial Times*, 2 October 1997.

65 Thanks to the participants in the meeting of late May 1998 in Valencia devoted to preparation of this Manifesto, and most especially to Jozef Celis, Raymond Pestiau and Jo van Cauwenberge (Belgium), Alain Fontaine (France), David Brubaker (USA), Claudio Jampaglia and Marinella Correggia (Italy), Thomas Kluge (Germany), and François Patenaude and Gabriel Regallet (Canada).

66 *Merryl Lynch Capital Markets*, 1 October 1997.

67 See the latest international report covering 16 countries, *Public/Private Partnerships of Water Supply throughout the World*, presented by G. P. Westerhoff and Malcolm Pirnie, Inc. to the World Congress of the International Water Resources Association (IWRA) in Madrid in 1997.

The IWRA published for the same occasion a 10-page brochure of comparative statistics on water prices in various countries and capital cities, on consumption per head of population, water distribution, and so on.

68 Jean-Jacques Donzier, 'La privatisation de la gestion de l'eau en France et en Angleterre: sous un même vocable, deux réalités bien différentes', in 'De l'eau pour demain', *Revue française de géoéconomie, op. cit.*, pp. 137–47.

69 Assemblée Nationale, *Le prix de l'eau: de l'exploitation à la maitrise*, report drafted by Ambroise Guelle, deputy, Paris, 1994.

70 The electricity sector, for its part, will have to repay £2.1 billion. Figures taken from *Le Devoir*, Montreal, 3 July 1997.

71 For a detailed analysis of water privatization in France and the UK, see Leo Paul Lauzor, François Patenaude and Martin Poirier, *La Privatisation de l'eau au Québec*, Part One, *Le cas des expériences en France et au Royaume-Uni*, Chaire d'études socio-économiques de l'UQAM, Montreal, April 1997.

72 We might also have considered the megacities of the developed countries as examples of the technological culture and techno-utopian vision of Westernized man. We have already referred to the implications of their growth for water, health and sanitation, and it is only lack of time that prevents us here from extending the analysis of town–country relations to other dimensions associated with the problem of technology lords.

73 The International Commission on Large Dams (ICOLD) defines a 'large dam' as one rising more than 15 metres from its base.

74 See the report *Large Dams, People and Environment Rights*, presented in Rome on 14 March as part of the Campaign for Reform of the World Bank. We are grateful to Francesco Martone, Liliana Cori and Antonio Tricarico, all active members of the Campaign, for their most helpful documentation.

75 In the French-speaking world, the specialist in this field is Patrick Lagadec, author of *Le risque technologique majeur*, Paris: Éd. Seuil, 1982.

76 According to Arundhati Roy (in her recent book *The Cost of Living*), the figures are largely understated and in India alone some 33 million people have been displaced as a result of the construction of three thousand large dams over the past fifty years.

77 For an analysis from a pro-desalination point of view, see Jacqueline Ribeiro, *Desalination Technology. Survey and Prospects*, Institute for Prospective Technological Studies, EUR 16434, Seville, 1996. An International Desalination Association is already in existence.

78 Published by Forum Umwelt Entwicklung, Bonn, 1998, with the support of the German Ministry of the Environment, Natural Conservation and Nuclear Reactor Safety.

79 The chapter title reads: 'Protection of the quality and supply of freshwater resources. Application of an integrated approach to the development, management and use of water resources'.
80 We know of numerous initiatives which for the moment remain more or less confined to small groups. One particularly noteworthy series of meetings was organized among representatives of the various world religions at Klingenthal in Alsace, on the initiative of Pax Christi. See *L'eau – Conclusions du 2e Symposium de Klingenthal*, Strasbourg: Pax Christi, 30 November 1997.

APPENDIX

International or World Organizations Specializing in Water-related Issues

- International Water Secretariat (Montreal)
- International Water Office (Paris, Limoges)
- International Water Resources Association (Southern Illinois University)
- Stockholm International Water Institute (Stockholm)
- World Water Council (Marseilles and Montreal)
- International Water Services Association (London)
- International Basin Organizations Network (Paris)
- International Association on Water Quality
- International Association on Water Law (Italy)
- International Water Supply Association
- Water Supply and Sanitation Collaborative Council (Geneva)
- International Commission on Irrigation and Drainage (New Delhi)
- International Association of Hydroelectricity (Lausanne)
- International Lake Environment Committee
- Académie de l'Eau (Paris)
- University Water Information Network (Carbondale, Illinois)
- Public Services International – Water Programme (Paris)
- International University Water Institute (Aix-en-Provence)
- World Resources Institute
- Worldwide Fund for Nature (Washington)
- Greenpeace
- Global Rivers Environment Education Network (Ann Arbor, Michigan)
- Third World Centre for Water Management (Mexico City)
- International Water Management Institute (Sri Lanka)
- World Conservation Union

United Nations Bodies Active in Relation to Water★

- UNCSD: United Nations Commission on Sustainable Development

- UNDP: United Nations Development Programme

- UNDP and World Bank: Water and Sanitation Programme

- UNEP: United Nations Environment Programme

- UNEP–GEMS: Global Environment Monitoring System

- FAO: Food and Agriculture Organization

- UNESCO: United Nations Educational, Scientific and Cultural Organization

- WMO: World Meteorological Organization

- UNICEF: United Nations Children's Fund

- WHO: World Health Organization

- The Economic and Social Commissions of the United Nations for the various continents (Europe, Latin America, Africa, Asia and the Pacific)

- UNU: United Nations University/International Network on Water, Environment and Health

And, in addition, two that are not part of the United Nations:

- World Bank

- IMF: International Monetary Fund

Source: Group of Lisbon

★ Nearly all the United Nations organizations have something to do with water. Those mentioned above play a special role in this area.

INDEX

The GLOBAL ISSUES Series

Already Available

In Preparation

Calestous Juma, *The New Genetic Divide: Biotechnology in the Age of Globalization*

John Madeley, *The New Agriculture: Towards Food for All*

Jeremy Seabrook, *The Future of Culture: Can Human Diversity Survive in a Globalized World?*

Harry Shutt, *A New Globalism: Alternatives to the Breakdown of World Order*

David Sogge, *Give and Take: Foreign Aid in the New Century*

Keith Suter, *Curbing Corporate Power: How Can We Control Transnational Corporations?*

Oscar Ugarteche, *A Level Playing Field: Changing the Rules of the Global Economy*

Nedd Willard, *The Drugs War: Is This the Solution?*

For full details of this list and Zed's other subject and general catalogues, please write to: The Marketing Department, Zed Books, 7 Cynthia Street, London N1 9JF, UK or email Sales@zedbooks.demon.co.uk

Visit our website at: http://www.zedbooks.demon.co.uk

PARTICIPATING ORGANIZATIONS

• **Both ENDS:** A service and advocacy organization which collaborates with environment and indigenous organizations, both in the South and in the North, with the aim of helping to create and sustain a vigilant and effective environmental movement.
Damrak 28-30, 1012 LJ Amsterdam, The Netherlands
Tel: +31 20 623 08 23   Fax: +31 20 620 80 49
Email: info@bothends.org
Website: www.bothends.org

• **Catholic Institute for International Relations (CIIR):** CIIR aims to contribute to the eradication of poverty through a programme that combines advocacy at national and international level with community-based development.
Unit 3 Canonbury Yard, 190a New North Road, London N1 7BJ, UK
Tel: +44 (0) 20 7354 0883   Fax: +44 (0) 20 7359 0017
Email: ciir@ciir.org
Website: www.ciir.org

• **Corner House:** The Corner House is a UK-based research and solidarity group working on social and environmental justice issues in North and South.
PO Box 3137, Station Road, Sturminster Newton, Dorset DT10 1YJ, UK
Tel: +44 (0)1258 473795   Fax: +44 (0)1258 473748
Email cornerhouse@gn.apc.org
Website: www.cornerhouse.icaap.org

• **Council on International and Public Affairs (CIPA):** CIPA is a human rights research, education and advocacy group, with a particular focus on economic and social rights in the USA and elsewhere around the world. Emphasis in recent years has been given to resistance to corporate domination.
777 United Nations Plaza, Suite 3C, New York, NY 10017, USA.
Tel: 212 972 9877   Fax: 212 972 9878
E-mail: cipany@igc.org
Website: www.cipa-apex.org

• **Dag Hammarskjöld Foundation:** The Dag Hammarskjöld Foundation, established 1962, organizes seminars and workshops on social, economic and cultural issues facing developing countries with a particular focus on alternative and innovative solutions. Results are published in its journal, *Development Dialogue*.

Övre Slottsgatan 2, 753 10 Uppsala, Sweden
Tel: 46 18 102772   Fax: 46 18 122072
e-mail: secretariat@dhf.uu.se
web site: www.dhf.uu.se

• **Development GAP:** The Development Group for Alternative Policies is a non-profit development resource organization working with popular organizations in the South and their Northern partners in support of a development that is truly sustainable and that advances social justice.

927 15th Street, NW - 4th Floor
Washington, DC 20005, USA
Tel: + 1-202-898-1566   Fax: +1 202-898-1612
E-mail: dgap@igc.org
Website: www.developmentgap.org

• **Focus on the Global South:** Focus is dedicated to regional and global policy analysis and advocacy work. It works to strengthen the capacity of organizations of the poor and marginalized people of the South and to better analyse and understand the impacts of the globalization process on their daily lives.

C/o CUSRI, Chulalongkorn University, Bangkok 10330, Thailand
Tel: +66 2 218 7363 Fax: + 66 2 255 9976
Email: Admin@focusweb.org
Website: www.focusweb.org

• **Inter Pares:** Inter Pares, a Canadian social justice organization, has been active since 1975 in building relationships with Third World development groups and providing support for community-based development programmes. Inter Pares is also involved in education and advocacy in Canada, promoting understanding about the causes, effects and solutions to poverty.

58 rue Arthur Street, Ottawa, Ontario, K1R 7B9 Canada
Tel: + 1 (613) 563-4801   Fax: + 1 (613) 594-4704

• **Third World Network:** TWN is an international network of groups and individuals involved in efforts to bring about a greater articulation of the needs and rights of peoples in the Third World; a fair distribution of the world's resources; and forms of development which are ecologically sustainable and fulfil human needs. Its international secretariat is based in Penang, Malaysia.

228 Macalister Road, 10400 Penang, Malaysia
Tel: +60-4-2266159   Fax: +60-4-2264505
Email: twnet@po.jaring.my
Website: www.twnside.org.sg

• **Third World Network–Africa:** TWN–Africa is engaged in research and advocacy on economic, environmental and gender issues. In relation to its current particular interest in globalization and Africa, its work focuses on trade and investment, the extractive sectors and gender and economic reform.

2 Ollenu Street, East Legon, P O Box AN19452, Accra-North, Ghana.
Tel: + 233 21 511189/503669/500419   Fax: + 233 21 51188
email: twnafrica@ghana.com

• **World Development Movement (WDM):** The World Development Movement campaigns to tackle the causes of poverty and injustice. It is a democratic membership movement that works with partners in the South to cancel unpayable debt and break the ties of IMF conditionality, for fairer trade and investment rules, and for strong international rules on multinationals.

25 Beehive Place, London SW9 7QR, UK
Tel: +44 20 7737 6215   Fax: +44 20 7274 8232
E-mail: wdm@wdm.org.uk
Website: www.wdm.org.uk

THIS BOOK IS AVAILABLE IN THE FOLLOWING COUNTRIES:

FIJI
University Book Centre
University of South Pacific
Suva

Tel: 679 313 900
Fax: 679 303 265

GHANA
EPP Book Services
P O Box TF 490
Trade Fair
Accra

Tel: 233 21 773087
Fax: 233 21 779099

MOZAMBIQUE
Sul Sensacoes
PO Box 2242,
Maputo

Tel: 258 1 421974
Fax: 258 1 423414

NEPAL
Everest Media Services
GPO Box 5443, Dillibazar
Putalisadak Chowk
Kathmandu

Tel: 977 1 416026
Fax: 977 1 250176

PAPUA NEW GUINEA
Unisearch PNG Pty Ltd
Box 320, University
National Capital District

Tel: 675 326 0130
Fax: 675 326 0127

RWANDA
Librairie Ikirezi
PO Box 443, Kigali

Tel/Fax: 250 71314

TANZANIA
TEMA Publishing Co Ltd
PO Box 63115
Dar Es Salaam

Tel: 255 51 113608
Fax: 255 51 110472

UGANDA
Aristoc Booklex Ltd
PO Box 5130, Kampala Road
Diamond Trust Building
Kampala

Tel/Fax: 256 41 254867

ZAMBIA
UNZA Press
PO Box 32379
Lusaka

Tel: 260 1 290409
Fax: 260 1 253952

ZIMBABWE
Weaver Press
P O Box A1922
Avondale, Harare

Tel: 263 4 308330
Fax: 263 4 339645